T0143079

Springer Series on Atomic, Optical, and Plasma Physics

Volume 51

The Springer Series on Atomic, Optical, and Plasma Physics covers in a comprehensive manner theory and experiment in the entire field of atoms and molecules and their interaction with electromagnetic radiation. Books in the series provide a rich source of new ideas and techniques with wide applications in fields such as chemistry, materials science, astrophysics, surface science, plasma technology, advanced optics, aeronomy, and engineering. Laser physics is a particular connecting theme that has provided much of the continuing impetus for new developments in the field, such as quantum computation and Bose-Einstein condensation. The purpose of the series is to cover the gap between standard undergraduate textbooks and the research literature with emphasis on the fundamental ideas, methods, techniques, and results in the field.

More information about this series at http://www.springer.com/series/411

Boris M. Smirnov

Atomic Particles and Atom Systems

Data for Properties of Atomic Objects and Processes

Second Edition

 Springer

Boris M. Smirnov
Joint Institute for High Temperatures
Russian Academy of Sciences
Moscow
Russia

ISSN 1615-5653   ISSN 2197-6791 (electronic)
Springer Series on Atomic, Optical, and Plasma Physics
ISBN 978-3-030-09222-1   ISBN 978-3-319-75405-5 (eBook)
https://doi.org/10.1007/978-3-319-75405-5

Originally published with the title "Reference Data on Atomic Physics and Atomic Processes"
1st edition: © Springer-Verlag Berlin Heidelberg 2008
2nd edition: © Springer International Publishing AG, part of Springer Nature 2018
Softcover re-print of the Hardcover 2nd edition 2018

Printed on acid-free paper

This Springer imprint is published by the registered company Springer International Publishing AG part of Springer Nature
The registered company address is: Gewerbestrasse 11, 6330 Cham, Switzerland

Preface

This book is devoted to atomic particles and processes involving atomic particles. It contains, on the one hand, numerical data for atoms, molecules, ions and processes with participation of electrons, ions, atoms and molecules. On the other hand, this book represents basic concepts and simple models for atomic particles and atomic processes. Numerical data are given in various forms including periodical tables of elements with filling by certain parameters of elements, atoms, molecules or processes, spectra of atoms, potential curves for electron terms of molecules, conversion factors for units used in formulas of atomic physics, tables and figures with certain atomic parameters. This information is joined with simple models for atomic particles and atomic processes.

Interaction inside atoms and between individual atomic particles is the basis of models described the nature of atomic particles and processes under consideration. Namely, electron terms of an atom follow from the exchange interaction of electrons inside an atom due to the Pauli exclusion principle, and the relation between exchange and spin–orbit interactions determines the atom coupling scheme, as well as atom quantum numbers. Comparison of the electrostatic and exchange interaction potential in a molecule with that of relativistic interactions gives molecular quantum numbers in accordance with the Hund's cases of momentum coupling. Interaction of atomic particles determines their dynamics, i.e. cross sections of their scattering in collisions. In turn, transport of atomic particles in gases and plasmas results from collisions of atomic particles, and, hence, transport coefficients are expressed through cross sections of collisions of atomic particles. Moving according to this scheme from atoms to molecules, from statics to dynamics, we add models and concepts with numerical data on each stage of consideration.

This book may be useful as an addition to contemporary books on atomic physics. Since the author collected and selected the content of this book in its practical work, it is directed to active specialists including advanced students who can use this material for the analysis of a certain physical situation.

Moscow, Russia   Boris M. Smirnov

Contents

Chapter 1
Introduction

This book is devoted to atomic particles, i.e., atoms, molecules, ions, and electrons. Properties of atomic particles, the character of their interaction, atomic collisions and transport phenomena in gases is the object of this book. As the second edition of the author book "Reference Data on Atomic Physics and Atomic Processes." (Berlin, Springer, 2008; 173p.) it gives numerical information about atomic particles and processes with their participation and accounts for that of contemporary reference books, in particular, for physical units and constants [1–4], and specific numerical parameters of atomic particles and atomic processes [2, 5–14]; along with this, some data are taken from contemporary reviews and original papers.

Selection of widely used information is the peculiarity of this reference book. As a result, we restrict ourselves by information for elements that gives the possibility to give it in the form of periodical tables of elements. The book contains 19 periodical tables with various atomic parameters. This allows us to increase the information density an take a simple key to find required data. Atomic spectra for the lowest atom states are represented in the book in the form of 28 Grotrian diagrams for atoms with the electron valence shell s, s^2, p^k. This information does not change during last tens years, and we use the Grotrian diagrams from [8, 15–18], and diagrams for the lowest atom states are taken from [19]. In addition, potential curves are represented for the ground and excited electron states of widespread molecules. Because this book intends for an active reader who makes some estimates and calculations, the transition between different units is required in various physical formulas. The book contains some tables with conversion factors for transition between units in corresponding formulas.

This book conserves general principles of the first edition which consists in selection the widely used information which is combined with simple models for the nature of atomic particles and processes with their participation. In particular, constructing an atom as a system of electrons located in the Coulomb field of the nucleus, we take as a basis the shell atom model. Within the framework of this model, electrons are distributed over electron shells which result from the one-electron

© Springer International Publishing AG, part of Springer Nature 2018
B. M. Smirnov, *Atomic Particles and Atom Systems*, Springer Series on Atomic, Optical, and Plasma Physics 51, https://doi.org/10.1007/978-3-319-75405-5_1

approximation where individual electrons are located in a self-conted field of the Coulomb center-nucleus and other electrons. Additional forms of interaction inside an atom include the exchange interaction of electrons which follows from the Pauli exclusion principle, and relativistic interactions inside an atoms where the main role for light atoms is spin-orbit interaction. Combination of these interactions determines various properties of atoms and molecules. For light atoms this leads to the Hund law that determines the sequence of atom levels owing to exchange interaction of electrons inside the atom for states with the same electron shell of valence electrons. Next, relation between the energies of exchange and spin-orbit interactions determines the character of coupling of electron momenta and, correspondingly, quantum numbers of atom states [20, 21]. This is demonstrated for certain atoms with using their measured parameters.

Continuing the analysis of interaction for two-atom systems, one can obtain new forms of interactions. At large distances between atomic particles compared with their sizes, one can divide this interaction in a long-range interaction which follows from interaction of momenta of atomic particles, and in a short-range interaction that results from exchange by electrons which belong to different atomic particles. Comparing these interactions and relativistic ones, one can obtain different cases for quantum numbers of a diatomic molecule which are known as the Hund cases of momentum coupling. From this electron properties follow for a diatomic molecule.

Along with properties of static atomic systems, interaction between atomic particles is of importance for dynamics of atomic particles and determines the character of particle scattering in their collisions. The hard sphere model is useful for the analysis of thermal collisions of atomic particles where atomic scattering is determined by a sharply varied interaction potential between colliding particles. This model is valid for some cases because a thermal energy is small compared to typical electron energies which are responsible for particle interaction. This model allows one to determine parameters of elastic atom-atom and ion-atom scattering in a simple manner, whereas the analysis of the resonant charge exchange process uses that the electron transition proceeds at large ion-atom distances compared with atomic sizes. The possibility to use these models is justifies by experimental data. Since transport phenomena in gases and plasmas, parameters of collisions of atomic particles determine the transport coefficients.

Thus, the problems under consideration are connected by a logical chain that starts from properties of atomic particles due to bound electrons and continues by interaction of atomic particles. The latter leads to certain properties of molecules and also to parameters of atomic collisions and transport phenomena in gases and plasmas. We attract experimental and theoretical data for parameters of certain problems on each step of this chain. Since this book intends for working specialists including advanced students, its content includes, on the one hand, numerical information and reference data for atomic particles and parameters of their dynamics, and, on the other hand, it considers the physics of atomic particles in the form of simple models, and reference data are combined with these models.

In this context it is necessary to explain the place of this reference book among other ones. Of course, it cannot compete with contemporary data banks.

In particular, NIST (National Institute of Science and Technology) is tuned on obtaining and propagation of "Standard Reference Data in Chemistry, Engineering, Fluids and Condensed Phases, Material Sciences, Mathematical and Computer Sciences and Physics" [22–25]. CODATA (Committee of Data for Science and Technology) represents contemporary data for various physical parameters [26]. HITRAN (High Resolution Transmission) molecular of absorption database [27–29] contains data for radiative properties of some molecules. Physical data of the above and other data banks represent rich information about atomic data, and its volume exceeds significantly that of this book.

Hence, the goal of this book is to give restricted information about atomic particles and processes, but it is frequently used. Of course, the question is what are frequently used data, and the author is based here on its experience as a physicist. Being guided by frequently used information, one can cut its volume, that simplify its search. Some help to this follows from some methods which include using of periodical tables as boxes for information.

Note that some physical parameters are known with a high accuracy. For example, the Planck constant [31] or speed of light [30] are known with the accuracy up to 9–10 sign. It may be of importance for the analysis of specific aspects of physics. But the accuracy with three-five signs is enough in the most of estimations in reality, and we are restricted by such accuracy. Next, the most part of information under consideration is known a long time. For example, used potential ionizations of not heavy atoms are the same as in the first half of 20 century. Therefore, new results with respect of the most data are not of interest in the plan of this book. One more remark is that some data often are known for a certain circle of elements or objects. Therefore we represent in this book a simple analysis and models for physics of objects under consideration. This can help to make simple estimations for objects where information is absent. But because the book is intended for working scientists, this part is not a textbook.

Thus, the main goal of this book is to extract general information about atomic particles and to apply this information to general models described properties and dynamics of interacting atomic particles. The author uses his practical experience in this field. In particular, a part of information about parameters of atomic objects is taken from Appendices to the author books [19, 32–38] devoted to certain aspects of atomic and plasma physics, so that the author collected this information as a user. Thus, starting from the nature of atomic objects and processes, we give detailed information about them. The book contains 77 tables with reference data, and 92 figures give certain information. This book may be used as an addition to existing books for various properties of atomic particles and their dynamics. Description of atomic objects on the basis of simple models is combined with information about objects and processes resulted from measurements. The author hopes that this material may be useful for active specialists in atom sciences.

Chapter 2
Elements of General Physics

Abstract Fundamental constants of general physics, physical units and conversional factors for units and physical formulas are represented. The system of physical units is constructedly combination of CGS (centimeter—gram—second) and ESU (electro-static units) or EMU (electromagnetic units) and the basic system of physical units is MKSA (meter—kilogram—second—ampere) or International System of Units (SI). Standard atomic weights and the natural occurrence of elements, abundance of stable isotopes in the Earth's crust and times of half-decay for long-lived radioactive isotopes are given.

2.1 Fundamental Physical Constants

Fundamental physical constants used in atomic and molecular physics are given in Table 2.1. Their accuracy is restricted by 6 signs that is enough for the most cases of the numerical analysis.

2.2 Elements and Isotopes

There are about 2700 stable and 2000 long-lived radioactive isotopes in the nature [39]. Stable isotopes relate to elements with the nuclear charge below 83, excluding technetium $_{43}Tc$ and promethium $_{61}Pm$, and also to elements with the nuclear charge in the range 90–93 (thorium, protactinium, uranium, neptunium). The diagram of Fig. 2.1 contains standard atomic masses for elements with taking into account their occurrence in the Earth crust [2, 40]. If stable isotopes of a given element are absent, the masses are given in square parentheses. These masses are given in atomic mass units (amu) where the unit is taken 1/12 mass of the carbon isotope $_{12}C$ from 1961 (see Table 2.1).

There is in diagram of Fig. 2.2 the occurrence of stable isotopes in the Earth crust. Diagram of Fig. 2.3 contain the lifetimes of stable of long-lived isotopes [2, 39–43]. Lifetimes of isotopes given in Fig. 2.3 is expressed in days (d) and years (y) and are

© Springer International Publishing AG, part of Springer Nature 2018   5
B. M. Smirnov, *Atomic Particles and Atom Systems*, Springer Series on Atomic,
Optical, and Plasma Physics 51, https://doi.org/10.1007/978-3-319-75405-5_2

given for isotopes whose lifetime exceed 2 h (0.08 d). In the case of transuranides included in Fig. 2.4 [43] these lifetimes may be smaller.

2.3 Physical Units

The unit system is a set of units through which various physical parameters are expressed. The basis of any mechanical system of units is the fact that the value of any dimensionality may be expressed through three dimensional units, the units of length, mass, time. The oldest system of units, the CGS-system of units, which basis are centimeter, gram and second [3], was introduced by British association for the Advancement of Science in 1874. The system of International Units [4] was adopted in 1960 (The conference des Poids et Mesure, Paris) and has as a basis meter (m), kilogram (kg), second (s). Other bases may be used for specific units.

In transferring from mechanics to other physics branches, additional units or assumption are required. In particular, below along with mechanical units, we deal with electric and magnetic units. Then the base SI units along with the above mechanical units contains the unit of an electric current ampere (A). Table 2.2 contains the SI units and their connection with the base units. Note that we express below the thermodynamic temperature through energy units.

In spreading the CGS-system of units to electrostatics, one can use the Coulomb law for the force F between two charges e_1 and e_2 that located at a distance r in a vacuum. This force is given by

$$F = \epsilon_o \frac{e_1 e_2}{r^2}$$

where ϵ_o is the vacuum permittivity. Defining a charge unit from this formula under the assumption $\epsilon_o = 1$, we construct in this manner the CGSE-system of units that describes physical parameters of mechanics and electrostatics. In the same way, in constructing the CGSM-system of units on the basis of the mechanical CGS-system of units the assumption is used that the vacuum magnetic conductivity is one $\mu_o = 1$. Because of units of electrostatic CGS (CGSE-system of units) and electromagnetic CGS (CGSM-system of units) are used, we below give some conversional between these systems and SI-units.

Along with the specific unit system, for atomic systems the system of atomic units (or Hartree atomic units) is of importance because parameters of atomic systems are expressed through atomic parameters. In construction the system of atomic units the fact is used that the parameter of any dimensionality may be built on the basis of three parameters of different dimensionality. As a basis of the system of atomic units the following three parameters are taken: the Planck constant $\hbar = 1.05457 \cdot 10^{-27}$ erg · s, the electron charge $e = 1.60218 \cdot 10^{-19}$ C and the electron mass $m_e = 9.10939 \cdot 10^{-28}$ g (we take then the vacuum permittivity and the magnetic conductivity of a vacuum to be one). The system of atomic units constructed on these parameters is given below in Table 2.3.

Tables 2.4, 2.5, 2.6, 2.7, 2.8, 2.9, 2.10, 2.11, 2.12, 2.13, 2.14 and 2.15 contain conversional factors between used units.

Physical units introduced in a honor of scientists

1. Unit of energy J (joule)—Joule J.P. (1818–1889, England)
2. Unit of temperature K (kelvin)—Kelvin, Lord (William Thomson, 1824–1907, Ireland)
3. Unit of frequency Hz (hertz)—Hertz H.R. (1857–1994, Germany)
4. Unit of power W (watt)—Watt J. (1736–1819, Scotland)
5. Unit of pressure Pa (pascal)—Pascal B. (1623–1662, France)
6. Unit of pressure torr (torrichelli)—Torrichelli E. (1608–1647, Italy)
7. Unit of force N (newton)—Newton I. (1643–1727, England)
8. Unit of charge C (coulomb)—Coulomb Ch.A. (1736–1806, France)
9. Unit of electric potential (voltage) V (volt)—Volta A.G.A.A. (1745–1827, Italy)
10. Unit of specific electric field strength Td (townsend)—Townsend J.S.E. (1868–1957, Ireland, England)
11. Unit of conductance S (siemens)—Siemens W. (1816–1892, England)
12. Unit of resistance Ω (ohm)—Ohm G.S. (1787–1854, Germany)
13. Unit of current strength A (ampere)—Ampere A.M. (1775–1836, France)
14. Unit of magnetic field strength Oe (oersted)—Oersted Ch. (1777–1851, Denmark)
15. Unit of magnetic flux Wb (weber)—Weber W.E. (1804–1891, Germany)
16. Unit of magnetic flux density Gs (gauss)—Gauss J.C.F. (1777–1855, Germany)
17. Unit of magnetic flux density T (tesla)—Tesla N. (1856–1943, Croatia, USA)
18. Unit of viscosity P (poise)—Poiseuille J.L.M. (1797–1869, France)
19. Unit of electric capacitance F (farad)—Faraday M. (1791–1867, England)
20. Unit of inductance H (henry)—Henry J. (1797–1878, USA)

2.4 Conversional Factors in Formulas of General Physics

Table 2.16 contains conversional factors in some physical formulas.
Explanations to Table 2.16.

1. The particle velocity is $v = \sqrt{2\varepsilon/m}$, where ε is the energy, m is the particle mass.
2. The average particle velocity is $v = \sqrt{8T/(\pi m)}$ with the Maxwell distribution function of particles on velocities; T is the temperature expressed in energetic units, m is the particle mass.
3. The particle energy is $\varepsilon = mv^2/2$, where m is the particle mass, v is the particle velocity.
4. The photon frequency is $\omega = \varepsilon/\hbar$, where ε is the photon energy.
5. The photon frequency is $\omega = 2\pi c/\lambda$, where λ is the wavelength.
6. The photon energy is $\varepsilon = 2\pi \hbar c/\lambda$.
7. The Larmor frequency is $\omega_H = eH/(mc)$ for a charged particle of a mass m in a magnetic field of a strength H.
8. The Larmor radius of a charged particle is $r_H = \sqrt{2\varepsilon/m}/\omega_H$, where ε is the energy of a charged particle, m is its mass, ω_H is the Larmor frequency.
9. The magnetic pressure $p_m = H^2/(8\pi)$.

Figures

Group / Period	I	II	Standard Atomic		
1	$_1$H 1.0079 Hydrogen	4.003 $_2$He Helium	**III**	**IV**	**V**
2	Li 6.941 Lithium	$_4$Be 9.012 Berillium	10.81 $_5$B Boron	12.011 $_6$C Carbon	14.007 $_7$N Nitrogen
3	$_{11}$Na 22.990 Sodium	$_{12}$Mg 24.305 Magnesium	26.982 $_{13}$Al Aluminium	28.086 $_{14}$Si Silicon	30.974 $_{15}$P Phosphorus
4	$_{19}$K 39.098 Potassium	$_{20}$Ca 40.08 Calcium	$_{21}$Sc 44.956 Scandium	$_{22}$Ti 47.867 Titanium	$_{23}$V 50.942 Vanadium
	63.546 $_{29}$Cu Copper	65.409 $_{30}$Zn Zinc	69.723 $_{31}$Ga Gallium	72.64 $_{32}$Ge Germanium	74.922 $_{33}$As Arsenic
5	$_{37}$Rb 85.468 Rubidium	$_{38}$Sr 87.62 Strontium	$_{39}$Y 88.906 Yttrium	$_{40}$Zr 91.224 Zirconium	$_{41}$Nb 92.906 Niobium
	107.87 $_{47}$Ag Silver	112.41 $_{48}$Cd Cadmium	114.82 $_{49}$In Indium	118.70 $_{50}$Sn Tin	121.76 $_{51}$Sb Antimony
6	$_{55}$Cs 132.90 Cesium	$_{56}$Ba 137.33 Barium	$_{57}$La 138.90 Lanthanum	$_{72}$Hf 178.49 Hafnium	$_{73}$Ta 180.95 Tantalum
	196.97 $_{79}$Au Gold	200.59 $_{80}$Hg Mercury	204.38 $_{81}$Tl Thallium	207.2 $_{82}$Pb Lead	208.98 $_{83}$Bi Bismuth
7	$_{87}$Fr [223.02] Francium	$_{88}$Ra [226.02] Radium	$_{89}$Ac [227.03] Actinium		

Actinides

$_{90}$Th 232.04 Thorium	$_{91}$Pa 231.04 Protactinium	$_{92}$U 238.03 Uranium	$_{93}$Np 237.05 Neptunium	$_{94}$Pu [244.06] Plutonium	$_{95}$Am [243.06] Americium
$_{96}$Cm [247.06] Curium	$_{97}$Bk [247.07] Berkelium	$_{98}$Cf [251.08] Californium	$_{99}$Es [252.08] Einsteinium	$_{100}$Fm [257.10] Fermium	$_{101}$Md [258.10] Mendelevium
$_{102}$No [259.10] Nobelium	$_{103}$Lr [262.11] Lawrencium	$_{104}$Rf [261.11] Rutherfordium	$_{105}$Db [262.11] Dubnium	$_{106}$Sg [266.12] Seaborgium	$_{107}$Bh [264.12] Bohrium

Fig. 2.1 Standard atomic weights of elements and occurrence of elements in the Earth's crust

Weights

VI	VII	VIII
15.999 $_8$O Oxygen	18.998 $_9$F Fluorine	20.180 $_{10}$Ne Neon
32.065 $_{16}$S Sulfur	35.453 $_{17}$Cl Chlorine	39.948 $_{18}$Ar Argon

Symbol — $_{42}$Mo   95.94
Atomic
number   Molibdenum
Element   Atomic weight

$_{24}$Cr 51.996 Chromium	$_{25}$Mn 54.938 Manganese	$_{26}$Fe 55.845 Iron	$_{27}$Co 58.933 Cobalt	$_{28}$Ni 58.69 Nickel
78.96 $_{34}$Se Selenium	79.904 $_{35}$Br Bromine	83.798 $_{36}$Kr Krypton		
$_{42}$Mo 95.94 Molibdenum	$_{43}$Tc [97.907] Technetium	$_{44}$Ru 101.07 Ruthenium	$_{45}$Rh 102.91 Rhodium	$_{46}$Pd 106.42 Palladium
127.60 $_{52}$Te Tellurium	126.90 $_{53}$I Iodine	131.29 $_{54}$Xe Xenon		
$_{74}$W 183.84 Tungsten	$_{75}$Re 186.21 Rhenium	$_{76}$Os 190.23 Osmium	$_{77}$Ir 192.22 Iridium	$_{78}$Pt 195.08 Platinum
[208.98] $_{84}$Po Polonium	[209.99] $_{85}$At Astatine	[222.02] $_{86}$Rn Radon		

Lantanides

$_{58}$Ce 140.12 Cerium	$_{59}$Pr 140.91 Praseodymium	$_{60}$Nd 144.24 Neodymium	$_{61}$Pm [144.91] Promethium	$_{62}$Sm 150.36 Samarium	$_{63}$Eu 151.96 Europium	$_{64}$Gd 157.25 Gadolinium
$_{65}$Tb 158.92 Terbium	$_{66}$Dy 162.50 Dysprosium	$_{67}$Ho 164.93 Holmium	$_{68}$Er 167.26 Erbium	$_{69}$Tm 168.93 Thulium	$_{70}$Yb 173.04 Ytterbium	$_{71}$Lu 174.97 Lutetium

Fig. 2.1 (continued)

Abundance of stable

Period	I	II	III	IV	V
1	$_1$H Hydrogen 1 - 99.985 2 - 0.015				
2	$_3$Li Lithium 6 - 7.5 7 - 92.5	$_4$Be Berillium 9 - 100	$_5$B Boron 10 - 19.9 11 - 80.1	$_6$C Carbon 12 - 98.9 13 - 1.1	$_7$N Nitrogen 14 - 99.63 15 - 0.37
3	$_{11}$Na Sodium 23 - 100	$_{12}$Mg Magnesium 24 - 79.0 25 - 10.0 26 - 11.0	$_{13}$Al Aluminium 27 - 100	$_{14}$Si Silicon 28 - 92.23 29 - 4.67 30 - 3.10	$_{15}$P Phosphorus 31 - 100
4	$_{19}$K Potassium 39 - 93.26 40 - 0.0117 41 - 6.73	$_{20}$Ca Calsium 40 - 96.94 42 - 0.647 43 - 0.135 44 - 2.09 46 - 0.004 48 - 0.187	$_{21}$Sc Scandium 45 - 100	$_{22}$Ti Titanium 46 - 8.0 47 - 7.3 48 - 73.8 49 - 5.5 50 - 5.4	$_{23}$V Vanadium 50 - 0.25 51 - 99.75
4	$_{29}$Cu Copper 63 - 69.17 65 - 30.83	$_{30}$Zn Zinc 64 - 48.6 66 - 27.9 67 - 4.1 68 - 19 70 - 0.6	$_{31}$Ga Gallium 69 - 60.1 71 - 39.9	$_{32}$Ge Germanium 70 - 20.2 72 - 27.2 73 - 7.8 74 - 36.1 76 - 8.7	$_{33}$As Arsenic 75 - 100
5	$_{37}$Rb Rubidium 85 - 72.17 87 - 27.83	$_{38}$Sr Strontium 84 - 0.56 86 - 9.86 87 - 7.00 88 - 82.58	$_{39}$Y Yttrium 89 - 100	$_{40}$Zr Zirconium 90 - 51.45 91 - 11.22 92 - 17.15 94 - 17.38 96 - 2.80	$_{41}$Nb Niobium 93 - 100
5	$_{47}$Ag Silver 107 - 51.84 109 - 48.16	$_{48}$Cd Cadmium 108 - 1.25 108 - 0.89 110 - 12.5 111 - 12.8 112 - 24.1 113 - 12.2 114 - 28.7 116 - 7.5	$_{49}$In Indium 113 - 4.3 115 - 95.7	$_{50}$Sn Tin 112 - 0.97 114 - 0.65 115 - 0.36 116 - 14.5 117 - 7.7 118 - 24.2 119 - 8.6 120 - 32.6 122 - 4.6 124 - 5.8	$_{51}$Sb Antimony 121 - 57 121 - 43
6	$_{55}$Cs Cesium 133 - 100	$_{56}$Ba Barium 130 - 0.106 132 - 0.101 134 - 2.42 135 - 6.59 136 - 7.9 137 - 11.2 138 - 71.7	$_{57}$La Lanthanum 138 - 0.09 139 - 99.91	$_{58}$Hf Hafnium 174 - 0.162 176 - 5.21 177 - 18.61 178 - 27.30 179 - 13.63 180 - 35.10	$_{73}$Ta Tantalum 180 - 0.012 181 - 99.988
6	$_{79}$Au Gold 197 - 100	$_{80}$Hg Mercury 196 - 0.1 198 - 10.0 199 - 16.8 200 - 23.1 201 - 13.2 202 - 29.8	$_{81}$Tl Thallium 204 - 6.8 203 - 29.52 205 - 70.48	$_{82}$Pb Lead 204 - 1.4 206 - 24.1 207 - 22.1 208 - 52.4	$_{83}$Bi Bismuth 208 - 100

Lantanides

$_{58}$Ce Cerium 136 - 0.19 138 - 0.25 140 - 88.5 142 - 11.1	$_{59}$Pr Praseo-dymium 141 - 100	$_{60}$Nd Neodymium 142 - 27.1 143 - 12.2 144 - 23.8 145 - 8.3 146 - 17.2 148 - 5.76 150 - 5.64	$_{61}$Pm Promethium	$_{62}$Sm Samarium 144 - 3.1 147 - 15.0 148 - 11.3 149 - 13.8 150 - 7.4 152 - 26.7 154 - 22.7

$_{63}$Eu Europium 151 - 48 153 - 52	$_{64}$Gd Gadolinium 152 - 0.20 154 - 2.18 155 - 14.8 156 - 20.5 157 - 15.6 158 - 24.8 160 - 21.9

Actinides

$_{90}$Th Thorium 232 - 100	$_{91}$Pa Protactinium 231 - 100	$_{92}$U Uranium 234 - 0.005 235 - 0.720 238 - 99.275

Fig. 2.2 Occurrence of stable isotopes in the Earth

isotopes

Legend: number of nucleons · abundance (%)

Symbol · Atomic number · Element

$_{77}$Ir — 191-37, 193-63 — Iridium

Group VIII

	number of nucleons · abundance (%)
$_2$He Helium	3 - 1.4 10^{-4}; 4 - 100

Groups VI, VII, VIII

Period 2

VI	VII	VIII
$_8$O Oxygen: 16 - 99.76, 17 - 0.04, 18 - 0.20	$_9$F Fluorine: 19 - 100	$_{10}$Ne Neon: 20 - 90.5, 21 - 0.27, 22 - 9.2

Period 3

VI	VII	VIII
$_{16}$S Sulfur: 32 - 95.0, 33 - 0.75, 34 - 4.2, 36 - 0.02	$_{17}$Cl Chlorine: 35 - 75.8, 37 - 24.2	$_{18}$Ar Argon: 36 - 0.337, 38 - 0.063, 40 - 99.60

Transition / Period 4

$_{24}$Cr Chromium	$_{25}$Mn Manganese	$_{26}$Fe Iron	$_{27}$Co Cobalt	$_{28}$Ni Nickel
50 - 4.35, 52 - 83.79, 53 - 9.50, 54 - 2.36	55 - 100	54 - 5.8, 56 - 92.0, 57 - 2.2, 58 - 0.28	59 - 100	58 - 68.27, 60 - 26.10, 61 - 1.13, 62 - 3.59, 64 - 0.91

VI	VII	VIII
$_{34}$Se Selenium: 74 - 0.9, 76 - 9.0, 77 - 7.6, 78 - 24.1, 80 - 50.2, 82 - 9.1	$_{37}$Br Bromine: 79 - 50.7, 81 - 49.3	$_{36}$Kr Krypton: 78 - 0.35, 80 - 2.25, 82 - 11.6, 83 - 11.5, 84 - 57.0, 86 - 17.3

Period 5

$_{42}$Mo Molibdenum	$_{43}$Tc Technetium	$_{44}$Ru Ruthenium	$_{45}$Rh Rhodium	$_{46}$Pd Palladium
92 - 14.8, 94 - 9.25, 95 - 15.9, 96 - 16.7, 97 - 9.55, 98 - 24.1, 100 - 9.63		96 - 5.5, 98 - 1.9, 99 - 12.7, 100 - 12.6, 101 - 17.0, 102 - 31.6, 104 - 18.7	103 - 100	102 - 1.02, 104 - 11.1, 105 - 22.3, 106 - 27.3, 108 - 26.5, 110 - 11.7

VI	VII	VIII
$_{52}$Te Tellurium: 120 - 0.096, 122 - 2.60, 123 - 0.91, 124 - 4.82, 125 - 7.14, 126 - 18.95, 128 - 31.69, 130 - 33.80	$_{53}$I Iodine: 127 - 100	$_{54}$Xe Xenon: 124 - 0.10, 126 - 0.09, 128 - 1.9, 129 - 26, 130 - 4.1, 131 - 21, 132 - 27, 134 - 10.4, 136 - 8.9

Period 6

$_{74}$W Tungsten	$_{75}$Re Rhenium	$_{76}$Os Osmium	$_{77}$Ir Iridium	$_{78}$Pt Platinum
180 - 0.1, 182 - 26.3, 183 - 14.3, 184 - 30.7, 186 - 28.6	185 - 37.4, 187 - 62.6	184 - 0.02, 186 - 1.6, 187 - 1.6, 188 - 13.3, 189 - 16, 190 - 26, 192 - 41	191 - 37, 193 - 63	190 - 0.01, 192 - 0.8, 194 - 33, 195 - 34, 196 - 25, 198 - 7.2

Lanthanides

$_{65}$Tb Terbium	$_{66}$Dy Dysprosium	$_{67}$Ho Holmium	$_{68}$Er Erbium	$_{69}$Tm Thulium	$_{70}$Yb Ytterbium	$_{71}$Lu Lutetium
159 - 100	156 - 0.06, 157 - 0.10, 160 - 2.3, 161 - 18.9, 162 - 25.5, 163 - 24.9, 164 - 28.2	165 - 100	162 - 0.14, 164 - 1.61, 166 - 33.6, 167 - 22.9, 168 - 26.8, 170 - 14.9	169 - 100	168 - 0.13, 170 - 3.1, 171 - 14.3, 172 - 21.9, 173 - 16.1, 174 - 32.0, 176 - 12.7	175 - 97.41, 176 - 2.59

Fig. 2.2 (continued)

Group / Period	I	II	III	IV	V
Long-lived (title over III–V)					
1	3 - 12.3 y ₁H Hydrogen	₂He Helium			
2	Li Lithium	7 - 53.4d 10 - 1.6·10⁶ y ₄Be Berillium	14 - 5730 y ₅B Boron	₆C Carbon	₇N Nitrogen
3	23 - 2.6y 24 - 0.63d ₁₁Na Sodium	28 - 0.88d ₁₂Mg Magnesium	26 - 7.2·10⁵ y ₁₃Al Aluminium	31 - 0.11d 32 - 330y ₁₄Si Silicon	32 -14.4d 33 - 25.3d ₁₅P Phosphorus
4 (a)	40 - 1.3 10⁹ y 42 - 0.52d 43 - 0.93d ₁₉K Potassium	41 - 1.4 10⁵ y 45 - 164d 47 - 4.54d 48 - 2.0 10¹⁶ y ₂₀Ca Calsium	43 - 0.19d 44 - 0.19d 46 - 84d 47 - 3.4d 48 - 1.83d ₂₁Sc Scandium	44 - 47y 45 - 0.13d ₂₂Ti Titanium	48 - 16d 49 - 0.9y 50 - 4.10¹⁶ y ₂₃V Vanadium
4 (b)	61 - 0.14d 64 - 0.53d 67 - 2.58d ₂₉Cu Copper	62 - 0.39d 65 - 244d 72 - 1.92d ₃₀Zn Zinc	66 - 0.40d 67 - 3.3d 72 - 0.59d 73 - 0.20d ₃₁Ga Gallium	66 - 0.10d 68 - 287d 69 - 1.62d 71 - 11.8d 77 - 0.47d ₃₂Ge Germanium	71 - 2.7d 77 - 1.6d 72 - 1.1d 73 - 80.3d 74 - 17.8d 76 - 1.1d ₃₃As Arsenic
5 (a)	81 - 0.19d 83 - 86.2d 84 - 33d 86 - 18.7d 87 - 4.8·10¹⁰ y ₃₇Rb Rubidium	82 - 25d 85 - 65d 83 - 1.4d 89 - 51d 90 - 29y 91 - 0.40d 92 - 0.11d ₃₈Sr Strontium	85 - 0.11d 87 - 3.3d 86 - 0.62d 88 - 107d 89 - 3.3d 90 - 2.7d 91 - 58.5d 92 - 0.15d 93 - 0.42d ₃₉Y Yttrium	86 - 0.69d 88 - 83d 89 - 3.3d 93 - 1.5 10⁶ y 95 - 64d ₄₀Zr 97 - 0.71d Zirconium	90 - 0.61d 91 - 1.0·10⁴ y 92 - 3.5·10⁷ y 94 - 2.0·10⁴ y 95 - 35d 96 - 0.97d ₄₁Nb Niobium
5 (b)	105 - 41.3d 111 - 7.5d 112 - 0.13d 113 - 0.22d ₄₇Ag Silver	107 - 0.27d 109 - 1.3y 113 - 9·10¹⁵ y 115 - 2.2d 117 - 0.10d ₄₈Cd Cadmium	109 - 0.18d 111 - 2.8d 115 - 4.4·10¹⁴ y ₄₉In Indium	110 - 0.17d 126 - 1·10⁵ y 113 - 115d 127 - 0.094d 121 - 1.1d 123 - 129d 125 - 9.6d ₅₀Sn Tin	117 - 0.12d 126 - 12.4d 119 - 1.59d 127 - 3.9d 120 - 5.8d 128 - 0.38d 122 - 2.7d 129 - 0.18d 124 - 60.2d 125 - 2.7y ₅₁Sb Antimony
6 (a)	127 - 0.26d 131 - 9.7d 129 - 1.34d 132 - 6.5d 134 - 2.06y 132 - 2.3 10⁶ y 136 - 13.2d ₅₅Cs 137 - 30y Cesium	128 - 2.4d 129 - 0.09d 131 - 12d 133 - 10.5d ₅₆Ba 140 - 12.8d Barium	132 - 0.20d 137 - 60000y 133 - 0.16d 138 - 1.3·10¹¹ y 135 - 0.79d 140 - 1.68 d 141 - 0.16 d ₅₇La Lanthanum	170 - 0.67d 173 - 1.0d 171 - 0.50d 174 - 2·10¹⁵ y 172 - 1.9y 175 - 70d 181 - 42.4d 182 - 9·10⁶ y ₇₂Hf 184 - 0.17d Hafnium	173 - 0.15d 177 - 2.4d 175 - 0.44d 178 - 0.09d 176 - 0.34d 179 - 1.82y 182 - 115d 183 - 5.1d ₇₃Ta 184 - 0.36d Tantalum
6 (b)	191 - 0.13d 197 - 2.7d 192 - 0.20d 199 - 3.14d 193 - 0.74d 194 - 1.65d 195 - 183d 196 - 6.2d ₇₉Au Gold	192 - 0.20d 203 - 46.6d 193 - 0.16d 194 - 260y 195 - 0.42d 197 - 2.7d ₈₀Hg Mercury	197 - 0.12d 202 - 12.2d 198 - 0.22d 204 - 3.8y 199 - 0.31d 200 - 1.1d 201 - 3.05d ₈₁Tl Thallium	198 - 0.10d 209 - 0.14d 200 - 0.90d 210 - 22.3y 201 - 0.39d 212 - 0.44d 202 - 3 10⁵ y 203 - 2.17d 205 - 1.4·10⁷ y ₈₂Pb Lead	203 - 0.49d 208 - 3.7·10⁵ y 204 - 0.47d 210 - 5.0d 205 - 15.3d 206 - 6.24d 207 - 38y ₈₃Bi Bismuth
7	₈₇Fr Francium	223 - 11.4d 224 - 3.7d 225 - 14.8d 226 - 1600y ₈₈Ra 228 - 5.8y Radium	224 - 0.12d 225 - 10d 226 - 1.2d 227 - 21.8y ₈₉Ac 228 - 0.25d Actinium		

Fig. 2.3 Lifetimes of long-lived radioactive isotopes: *s*—second, *m*—minute, *h*—hour, *d*—day, *y*—year

isotopes

Legend:

Number of nucleons — Time of half-decay

Symbol Mo · Atomic number 42 · Element Molibdenum · $90 - 0.24d$, $91 - 3500y$, $99 - 2.8d$

VI

$_8O$ — Oxygen

$_{16}S$ — Sulfur
35 - 87.2d; 38 - 0.12d

$_{24}Cr$ — Chromium
48 - 0.96d; 51 - 27.7d

$_{34}Se$ — Selenium
72 - 8.4d; 73 - 0.30d; 75 - 120d; 79 - 65000y

$_{42}Mo$ — Molibdenum
90 - 0.24d; 91 - 3500y; 99 - 2.8d

$_{52}Te$ — Tellurium
116 - 0.10d; 118 - 6.0d; 119 - 0.67d; 121 - 17d; 123 - 10^{13} y; 127 - 0.39d; 132 - 3.26d

$_{74}W$ — Tungsten
176 - 0.10d; 177 - 0.09d; 178 - 22d; 181 - 121d; 185 - 75d; 187 - 1.0d; 188 - 69d

$_{84}Po$ — Polonium
204 - 0.15d; 206 - 8.8d; 207 - 0.24d; 208 - 2.9y; 209 - 102y; 210 - 138.4d

VII

$_9F$ — Fluorine

$_{17}Cl$ — Chlorine
36 - $3.0 \cdot 10^5$ y

$_{25}Mn$ — Manganese
52 - 5.6d; 53 - $3.7 \cdot 10^6$ y; 54 - 312d; 56 - 0.11d

$_{35}Br$ — Bromine
76 - 0.68d; 77 - 2.4d; 82 - 1.47d; 83 - 0.10d

$_{43}Tc$ — Technetium
93 - 0.12d; 94 - 0.20d; 95 - 0.83d; 96 - 4.3d; 97 - $2.6 \cdot 10^6$ y; 98 - $4.2 \cdot 10^6$ y; 99 - $2.1 \cdot 10^5$ y

$_{53}I$ — Iodine
121 - 0.09d; 123 - 0.55d; 124 - 4.2d; 125 - 60d; 126 - 13d; 129 - $1.6 \; 10^7$ y; 130 - 0.52d; 131 - 8.0d; 132 - 0.10d; 133 - 0.87d; 135 - 0.28d

$_{75}Re$ — Rhenium
181 - 0.83d; 182 - 2.67d; 183 - 70d; 184 - 38d; 186 - 3.8d; 187 - $5 \; 10^{10}$ y; 188 - 0.71d

$_{85}At$ — Astatine
209 - 0.22d; 210 - 0.34d; 211 - 0.30d

VIII

$_{10}Ne$ — Neon

$_{18}Ar$ — Argon
37 - 35d; 39 - 269y; 42 - 33y

$_{26}Fe$ — Iron
52 - 0.35d; 55 - 2.7y; 59 - 44.5d; 60 - 3.0 10^5 y

$_{27}Co$ — Cobalt
55 - 0.73d; 56 - 79d; 57 - 271d; 58 - 71d; 60 - 5.3y

$_{28}Ni$ — Nickel
56 - 6.1d; 57 - 1.5d; 59 - 7.5 10^4 y; 63 - 100y; 65 - 0.10d; 66 - 2.3d

$_{36}Kr$ — Krypton
76 - 0.62d; 79 - 1.46d; 81 - $2 \cdot 10^5$ y; 85 - 10.7y; 88 - 0.12d

$_{44}Ru$ — Ruthenium
97 - 2.9d; 103 - 39.3d; 105 - 0.18d; 106 - 1.2y

$_{45}Rh$ — Rhodium
99 - 16d; 100 - 0.88d; 101 - 3.3y; 102 - 2.9y; 105 - 1.5d

$_{46}Pd$ — Palladium
100 - 3.6d; 101 - 0.35d; 103 - 17d; 107 - $6.5 \cdot 10^6$ y; 109 - 0.56d

$_{54}Xe$ — Xenon
122 - 0.84d; 123 - 0.09d; 125 - 0.71d; 127 - 36.4d; 133 - 5.29d; 135 - 0.38d

$_{76}Os$ — Osmium
182 - 0.92d; 183 - 0.54d; 185 - 94d; 186 - $2 \cdot 10^{15}$ y; 191 - 15.4d; 193 - 1.27d; 194 - 6.0y

$_{77}Ir$ — Iridium
184 - 0.12d; 185 - 0.58d; 187 - 0.44d; 188 - 1.7d; 189 - 13.2d; 190 - 12d; 192 - 73.8d; 194 - 0.80d; 195 - 0.10d

$_{78}Pt$ — Platinum
186 - 0.08d; 187 - 0.10d; 188 - 10.2d; 189 - 0.45d; 190 - $6 \cdot 10^{11}$ y; 191 - 2.9d; 193 - 50y; 197 - 0.75d; 200 - 0.52d

$_{86}Rn$ — Radon
210 - 0.10d; 211 - 0.61d; 222 - 3.82d

Fig. 2.3 (continued)

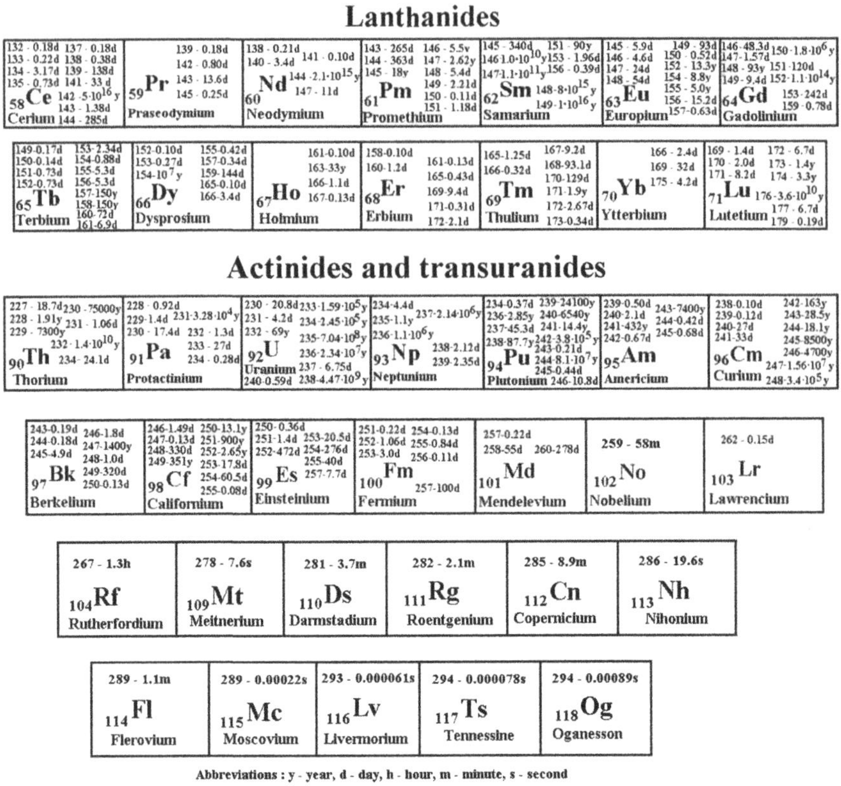

Lanthanides

58 Ce — Cerium: 132 - 0.18d, 133 - 0.22d, 134 - 3.17d, 135 - 0.73d, 137 - 0.18d, 138 - 0.38d, 139 - 138d, 141 - 33 d, 142 - $5 \cdot 10^{16}$ y, 143 - 1.38d, 144 - 285d

59 Pr — Praseodymium: 139 - 0.18d, 142 - 0.80d, 143 - 13.6d, 145 - 0.25d

60 Nd — Neodymium: 138 - 0.21d, 140 - 3.4d, 141 - 0.10d, 144 - $2.1 \cdot 10^{15}$ y, 147 - 11d

61 Pm — Promethium: 143 - 265d, 144 - 363d, 145 - 18y, 146 - 5.5v, 147 - 2.62y, 148 - 5.4d, 149 - 2.21d, 150 - 0.11d, 151 - 1.18d

62 Sm — Samarium: 145 - 340d, 146 $1.0 \cdot 10^{10}$ y, 147 $1.1 \cdot 10^{11}$ y, 148 - $8 \cdot 10^{15}$ y, 149 - $1 \cdot 10^{16}$ y, 151 - 90y, 153 - 1.96d, 156 - 0.39d

63 Eu — Europium: 145 - 5.9d, 146 - 4.6d, 147 - 24d, 148 - 54d, 155 - 5.0y, 156 - 15.2d, 157 - 0.63d

64 Gd — Gadolinium: 149 - 93d, 150 - 0.52d, 152 - 13.3y, 154 - 8.8y, 146 - 48.3d, 147 - 1.57d, 148 - 93y, 149 - 9.4d, 150 - $1.8 \cdot 10^6$ y, 151 - 120d, 152 - $1.1 \cdot 10^{14}$ y, 153 - 242d, 159 - 0.78d

65 Tb — Terbium: 149 - 0.17d, 150 - 0.14d, 151 - 0.73d, 152 - 0.73d, 153 - 2.34d, 154 - 0.88d, 155 - 5.3d, 156 - 5.3d, 157 - 150y, 158 - 150y, 160 - 72d, 161 - 6.9d

66 Dy — Dysprosium: 152 - 0.10d, 153 - 0.27d, 154 - 10^7 y, 155 - 0.42d, 157 - 0.34d, 159 - 144d, 165 - 0.10d, 166 - 3.4d

67 Ho — Holmium: 161 - 0.10d, 163 - 33y, 166 - 1.1d, 167 - 0.13d

68 Er — Erbium: 158 - 0.10d, 160 - 1.2d, 161 - 0.13d, 165 - 0.43d, 169 - 9.4d, 171 - 0.31d, 172 - 2.1d

69 Tm — Thulium: 165 - 1.25d, 166 - 0.32d, 167 - 9.2d, 168 - 93.1d, 170 - 129d, 171 - 1.9y, 172 - 2.67d, 173 - 0.34d

70 Yb — Ytterbium: 166 - 2.4d, 169 - 32d, 175 - 4.2d

71 Lu — Lutetium: 169 - 1.4d, 170 - 2.0d, 171 - 8.2d, 172 - 6.7d, 173 - 1.4y, 174 - 3.3y, 176 - $3.6 \cdot 10^{10}$ y, 177 - 6.7d, 179 - 0.19d

Actinides and transuranides

90 Th — Thorium: 227 - 18.7d, 228 - 1.9ly, 229 - 7300y, 230 - 75000y, 231 - 1.06d, 232 - $1.4 \cdot 10^{10}$ y, 234 - 24.1d

91 Pa — Protactinium: 228 - 0.92d, 229 - 1.4d, 230 - 17.4d, 231 - $3.28 \cdot 10^4$ y, 232 - 1.3d, 233 - 27d, 234 - 0.28d

92 U — Uranium: 230 - 20.8d, 231 - 4.2d, 232 - 69y, 233 - $1.59 \cdot 10^5$ y, 234 - $2.45 \cdot 10^5$ y, 235 - $7.04 \cdot 10^8$ y, 236 - $2.34 \cdot 10^7$ y, 237 - 6.75d, 238 - $4.47 \cdot 10^9$ y

93 Np — Neptunium: 234 - 4.4d, 235 - 1.1y, 236 - $1.1 \cdot 10^6$ y, 237 - $2.14 \cdot 10^6$ y, 238 - 2.12d, 239 - 2.35d

94 Pu — Plutonium: 234 - 0.37d, 236 - 2.85y, 237 - 45.3d, 238 - 87.7y, 239 - 24100y, 240 - 6540y, 241 - 14.4y, 242 - $3.8 \cdot 10^5$ y, 243 - 0.21d, 244 - $8.1 \cdot 10^7$ y, 245 - 10.5d, 246 - 10.8d

95 Am — Americium: 239 - 0.50d, 240 - 2.1d, 241 - 432y, 242 - 0.67d, 243 - 7400y, 244 - 0.42d, 245 - 0.68d

96 Cm — Curium: 238 - 0.10d, 239 - 0.12d, 240 - 27d, 241 - 33d, 242 - 163y, 243 - 28.5y, 244 - 18.1y, 245 - 8500y, 246 - 4700y, 247 - $1.56 \cdot 10^7$ y, 248 - $3.4 \cdot 10^5$ y

97 Bk — Berkelium: 243 - 0.19d, 244 - 0.18d, 245 - 4.9d, 246 - 1.8d, 247 - 1400y, 248 - 1.0d, 249 - 320d, 250 - 0.13d

98 Cf — Californium: 246 - 1.49d, 247 - 0.13d, 248 - 330d, 249 - 351y, 250 - 13.1y, 251 - 900y, 252 - 2.65y, 253 - 17.8d, 254 - 60.5d, 255 - 0.08d

99 Es — Einsteinium: 250 - 0.36d, 251 - 1.4d, 252 - 472d, 253 - 20.5d, 254 - 276d, 255 - 40d, 256 - 0.11d, 257 - 7.7d

100 Fm — Fermium: 251 - 0.22d, 252 - 1.06d, 253 - 3.0d, 254 - 0.13d, 255 - 0.84d, 256 - 0.11d, 257 - 100d

101 Md — Mendelevium: 257 - 0.22d, 258 - 55d, 260 - 278d

102 No — Nobelium: 259 - 58m

103 Lr — Lawrencium: 262 - 0.15d

104 Rf — Rutherfordium: 267 - 1.3h

109 Mt — Meitnerium: 278 - 7.6s

110 Ds — Darmstadium: 281 - 3.7m

111 Rg — Roentgenium: 282 - 2.1m

112 Cn — Copernicium: 285 - 8.9m

113 Nh — Nihonium: 286 - 19.6s

114 Fl — Flerovium: 289 - 1.1m

115 Mc — Moscovium: 289 - 0.00022s

116 Lv — Livermorium: 293 - 0.000061s

117 Ts — Tennessine: 294 - 0.000078s

118 Og — Oganesson: 294 - 0.00089s

Abbreviations: y - year, d - day, h - hour, m - minute, s - second

Fig. 2.4 Lifetimes of radioactive isotopes of lathanides and transuranides: *s*—second, *m*—minute, *h*—hour, *d*—day, *y*—year

Tables

Table 2.1 Fundamental physical constants

Electron mass	$m_e = 9.10938 \cdot 10^{-28}$ g
Proton mass	$m_p = 1.67262 \cdot 10^{-24}$ g
Atomic unit of mass	$m_a = \frac{1}{12}m(^{12}C) = 1.660539 \cdot 10^{-24}$ g
Ratio of proton and electron masses	$m_p/m_e = 1836.15$
Ratio of atomic and electron masses	$m_a/m_e = 1822.89$
Electron charge	$e = 1.602177 \cdot 10^{-19}$ C $= 4.803204 \cdot 10^{-10}$ e.s.u.
	$e^2 = 2.3071 \cdot 10^{-19}$ erg $\cdot$ cm $= 14.3996$ eV Å
Planck constant	$h = 6.62619 \cdot 10^{-27}$ erg s
	$\hbar = 1.054572 \cdot 10^{-27}$ erg s
Light velocity	$c = 2.99792458 \cdot 10^{10}$ cm/s, $m_e c^2 = 510.98$ keV
Fine-structure constant	$\alpha = e^2/(\hbar \cdot c) = 0.07295$
Inverse fine-structure constant	$1/\alpha = \hbar \cdot c/e^2 = 137.03599$
Bohr radius	$a_o = \hbar^2/(m_e \cdot e^2) = 0.5291772$ Å
Rydberg constant	$R = m_e e^4/(2\hbar^2) = 13.60569$ eV $= 2.17987$ $\cdot 10^{-18}$ J
Bohr magneton	$\mu_B = e\hbar/(2m_e c) = 9.27401 \cdot 10^{-24}$ J/T $=$ $9.27401 \cdot 10^{-21}$ erg/Gs
Avogadro number	$N_A = 6.02214 \cdot 10^{23}$ mol^{-1}
Stephan-Boltzmann constant	$\sigma = \pi^2/(60\hbar^3 c^2)$
Molar volume	$R = 22.414$ l/mol
Loschmidt number	$L = N_A/R = 2.6867 \cdot 10^{19}$ cm^{-3}
Faraday constant	$F = N_A e = 96485.3$ C/mol

Table 2.2 Basic SI units [1, 2, 4]

Quantity	Name	Symbol	Connection with base units
Frequency	hertz	Hz	$1/s$
Force	newton	N	$m\,kg/s^2$
Pressure	pascal	Pa	$kg/(m\,s^2)$
Energy	joule	J	$m^2\,kg/s^2$
Power	watt	W	$m^2\,kg/s^3$
Charge	coulomb	C	$A\,s$
Electric potential (voltage)	volt	V	W/A
Electric capacitance	farad	F	$A^2\,s^4/(m^2\,kg)$
Electric resistance	ohm	Ω	$m^2\,kg/(A^2\,s^3)$
Conductance	siemens	S	$A^2\,s^3/(m^2\,kg)$
Inductance	henry	H	$m^2\,kg/(A^2\,s^2)$
Magnetic flux	weber	Wb	$m^2\,kg/(A\,s^2)$
Magnetic flux density	tesla	T	$kg/(A\,s^2)$

Table 2.3 System of atomic units

Parameter	Symbol, formula	Value
Length	$a_o = \hbar^2/(me^2)$	$5.2918 \cdot 10^{-9}$ cm
Velocity	$v_o = e^2/\hbar$	$2.1877 \cdot 10^8$ cm/s
Time	$\tau_o = \hbar^3/(me^4)$	$2.4189 \cdot 10^{-17}$ s
Frequency	$\nu_o = me^4/\hbar^3$	$4.1341 \cdot 10^{16}$ s^{-1}
Energy	$\varepsilon_o = me^4/\hbar^2$	$27.2114\,eV = 4.3598 \cdot 10^{-18}$ J
Power	$\varepsilon_o/\tau = m^2e^8/\hbar^5$	0.18024 W
Electric voltage	$\varphi_o = me^3/\hbar^2$	27.2114 V
Electric field strength	$E_o = me^5/\hbar^4$	$5.1422 \cdot 10^9$ V/cm
Momentum	$p_o = me^2/\hbar$	$1.9929 \cdot 10^{-19}$ g cm/s
Number density	$N_o = a_o^{-3}$	$6.7483 \cdot 10^{24}$ cm^{-3}
Volume	$V_o = a_o^3$	$1.4818 \cdot 10^{-25}$ cm^3 = 0.089240 cm^3/mol
Square, cross section	$\sigma_o = a_o^2$	$2.8003 \cdot 10^{-21}$ cm^2
Rate constant	$k_o = v_o a_o^2 = \hbar^3/(m^2e^2)$	$6.126 \cdot 10^{-9}$ cm^3/s
Three body rate constant	$K_o = v_o a_o^5 = \hbar^9/(m^5e^8)$	$9.078 \cdot 10^{-34}$ cm^6/s
Dipole moment	$ea_o = \hbar^2/(me)$	$2.5418 \cdot 10^{-18}$ esu = 2.5418 D
Magnetic moment	$\hbar^2/(me) = 2\mu_B/\alpha$	$2.5418 \cdot 10^{-18}$ erg/Gs = $2.5418 \cdot 10^{-21}$ J/T
Electric current	$I = e/\tau = me^5/\hbar^3$	$6.6236 \cdot 10^{-3}$ A
Flux	$j_o = N_o v_o = m^3e^8/\hbar^7$	$1.476 \cdot 10^{33}$ cm^{-2} s^{-1}
Electric current density	$i_o = eN_o v_o = m^3e^9/\hbar^7$	$2.3653 \cdot 10^{14}$ A/cm^2
Energy flux	$J = \varepsilon_o N_o v_o = m^4e^{12}/\hbar^9$	$6.436 \cdot 10^{15}$ W/cm^2

Table 2.4 Conversional factors for units of energy

	1 J	1 erg	1 eV	1 K	1 cm^{-1}	1 MHz	1 kcal/mol	1 kJ/mol
1 J	1	10^7	$6.2415 \cdot 10^{18}$	$7.2429 \cdot 10^{22}$	$5.0341 \cdot 10^{22}$	$9.482 \cdot 10^{27}$	$1.4393 \cdot 10^{20}$	$6.0221 \cdot 10^{20}$
1 erg	10^{-7}	1	$6.2415 \cdot 10^{11}$	$7.2429 \cdot 10^{15}$	$5.0341 \cdot 10^{15}$	$9.482 \cdot 10^{20}$	$1.4393 \cdot 10^{13}$	$6.0221 \cdot 10^{13}$
1 eV	$1.6022 \cdot 10^{-19}$	$1.6022 \cdot 10^{-12}$	1	11604	8065.5	$1.519 \cdot 10^9$	23.045	96.485
1 K	$1.3807 \cdot 10^{-23}$	$1.3807 \cdot 10^{-16}$	$8.6174 \cdot 10^{-5}$	1	0.69504	$1.309 \cdot 10^5$	$1.9872 \cdot 10^{-3}$	$8.3145 \cdot 10^{-3}$
1 cm^{-1}	$1.9864 \cdot 10^{-23}$	$1.9864 \cdot 10^{-16}$	$1.2398 \cdot 10^{-4}$	1.4388	1	$1.8837 \cdot 10^5$	$2.8591 \cdot 10^{-3}$	$1.1963 \cdot 10^{-2}$
1 MHz	$1.055 \cdot 10^{-28}$	$1.055 \cdot 10^{-21}$	$6.582 \cdot 10^{-10}$	$7.638 \cdot 10^{-6}$	$5.309 \cdot 10^{-6}$	1	$1.8568 \cdot 10^{-3}$	$4.438 \cdot 10^{-4}$
1 kcal/mol	$6.9477 \cdot 10^{-21}$	$6.9477 \cdot 10^{-28}$	$4.3364 \cdot 10^{-2}$	503.22	349.76	538.6	1	4.184
1 kJ/mol	$1.6605 \cdot 10^{-21}$	$1.6605 \cdot 10^{-28}$	$1.0364 \cdot 10^{-2}$	120.27	83.594	2253	0.23901	1

Table 2.5 Conversional factors for units of pressure

	$1 Pa = 1 N/m^2$	$1 dyn/cm^2$	1 Torr	1 atm[a]	1 at[b]	1 bar
$1 Pa = 1 N/m^2$	1	10	$7.5001 \cdot 10^{-3}$	$9.8693 \cdot 10^{-6}$	$1.0197 \cdot 10^{-5}$	10^{-5}
$1 dyn/cm^2$	0.1	1	$7.5001 \cdot 10^{-4}$	$9.8693 \cdot 10^{-7}$	$1.0197 \cdot 10^{-6}$	10^{-6}
1 Torr	133.332	1333.32	1	$1.3158 \cdot 10^{-3}$	$1.3595 \cdot 10^{-3}$	$1.33332 \cdot 10^{-3}$
1 atm[a]	$1.01325 \cdot 10^5$	$1.01325 \cdot 10^6$	760	1	1.01332	1.01325
1 at[b]	$9.80665 \cdot 10^4$	$9.80665 \cdot 10^5$	735.56	0.96785	1	0.980665
1 bar	10^5	10^6	750.01	0.98693	1.0197	1

[a] atm—physical atmosphere
[b] at $= kg/cm^2$—technical atmosphere

Table 2.6 Units of electric charge

	1 e	1 CGSE	1 C
1 e	1	$4.8032 \cdot 10^{-10}$	$1.60218 \cdot 10^{-19}$
1 CGSE	$2.0819 \cdot 10^9$	1	$3.33564 \cdot 10^{-10}$
1 C	$6.2415 \cdot 10^{18}$	$2.99792 \cdot 10^9$	1

Table 2.7 Units of electric voltage

	1 V	1 CGSE	1 C/m
1 V	1	$3.33564 \cdot 10^{-3}$	$1.113 \cdot 10^{-10}$
1 CGSE	299.792	1	$3.3364 \cdot 10^{-8}$
1 C/m	$8.9875 \cdot 10^9$	$2.99792 \cdot 10^7$	1

Table 2.8 Units of electric field strength

	1 V/cm	1 CGSE	$1 C/m^2$
1 V/cm	1	$3.33564 \cdot 10^{-3}$	$1.113 \cdot 10^{-8}$
1 CGSE	299.792	1	$3.3364 \cdot 10^{-6}$
$1 C/m^2$	$8.9875 \cdot 10^7$	$2.99792 \cdot 10^5$	1

Table 2.9 Units of specific electric field strength

	1 Td [a]	1 V/(cm Torr)
1 Td [a]	1	2.829
1 V/(cm Torr)	0.3535	1

[a] $1 Td = 1 \cdot 10^{-17} V cm^2$

Table 2.10 Units of conductivity

	S/m	$1/(\Omega cm)$	1/s
S/m	1	0.01	$1.11265 \cdot 10^{-14}$
$1/(\Omega cm)$	100	1	$1.11265 \cdot 10^{-12}$
1/s	$8.98755 \cdot 10^{13}$	$8.98755 \cdot 10^{11}$	1

Table 2.11 Units of electric resistance

	1 Ω	1 CGSE	1 CGSM
1 Ω	1	$1.11265 \cdot 10^{-12}$	10^9
1 CGSE	$8.98755 \cdot 10^{11}$	1	$8.98755 \cdot 10^{20}$
1 CGSM	10^{-9}	$1.11265 \cdot 10^{-21}$	1

Table 2.12 Units of current density

	1 e/(cm^2 s)	1 CGSE	1 A/m^2
1 e/(cm^2 s)	1	2.99792 $4.8032 \cdot 10^{-10}$	$1.60218 \cdot 10^{-15}$
1 CGSE	$2.0819 \cdot 10^9$	1	$3.3356 \cdot 10^{-6}$
1 A/m^2	$6.2415 \cdot 10^{14}$	$2.9979 \cdot 10^5$	1

Table 2.13 Units of magnetic field strength

	1 Oe	1 CGSE	1 A/m
1 Oe	1	$2.99792 \cdot 10^{10}$	79.5775
1 CGSE	$3.33564 \cdot 10^{-11}$	1	$2.65442 \cdot 10^{-9}$
1 A/m	0.012566	$1.11265 \cdot 10^{-21}$	1

Table 2.14 Units of magnetic induction

	1 CGSE	1 T = 1 Wb/m^2	1 Gs
1 CGSE	1	$2.99792 \cdot 10^6$	$2.99792 \cdot 10^{10}$
1 T = 1 Wb/m^2	$3.33564 \cdot 10^{-7}$	1	10^4
1 Gs	$3.33564 \cdot 10^{-11}$	10^{-4}	1

Table 2.15 Units of viscocity

	1 CGSE = g/(cm s)	1 P (poise)	1 Pa s
1 CGSE = g/(cm s)	1	1	0.1
1 P (poise)	1	1	0.1
1 Pa s	10	10	1

Table 2.16 Conversional factors for formulas involving atomic particles

Number	Formula[a]	Factor C	Units used
1	$v = C\sqrt{\varepsilon/m}$	$5.931 \cdot 10^7$ cm/s	ε in eV, m in e.m.u.[a]
		$1.389 \cdot 10^6$ cm/s	ε in eV, m in a.m.u.[a]
		$5.506 \cdot 10^5$ cm/s	ε in K, m in e.m.u.
		$1.289 \cdot 10^4$ cm/s	ε in K, m in a.m.u.
2	$v = C\sqrt{T/m}$	$1.567 \cdot 10^6$ cm/s	T in eV, m in a.m.u.
		$1.455 \cdot 10^4$ cm/s	T in K, m in a.m.u.
		$6.212 \cdot 10^5$ cm/s	T in K, m in a.m.u.
		$6.692 \cdot 10^7$ cm/s	T in eV, m in e.m.u.
3	$\varepsilon = Cv^2$	$3.299 \cdot 10^{-12}$ K	v in cm/s, m in e.m.u.
		$6.014 \cdot 10^{-9}$ K	v in cm/s, m in a.m.u.
		$2.843 \cdot 10^{-16}$ eV	v in cm/s, m in e.m.u.
		$5.182 \cdot 10^{-13}$ eV	v in cm/s, m in a.m.u.
4	$\omega = C\varepsilon$	$1.519 \cdot 10^{15}$ s^{-1}	ε in eV
		$1.309 \cdot 10^{11}$ s^{-1}	ε in K
5	$\omega = C/\lambda$	$1.884 \cdot 10^{15}$ s^{-1}	λ in μm
6	$\varepsilon = C/\lambda$	1.2398 eV	λ in μm
7	$\omega_H = CH/m$	$1.759 \cdot 10^7$ s^{-1}	H in Gs, m in e.m.u.
		9655 s^{-1}	H in Gs, m in a.m.u.
8	$r_H = C\sqrt{\varepsilon m}/H$	3.372 cm	ε in eV, m in e.m.u., H in Gs
		143.9 cm	ε in eV, m in a.m.u., H in Gs
		$3.128 \cdot 10^{-2}$ cm	ε in K, m in e.m.u., H in Gs
		1.336 cm	ε in K, m in a.m.u., H in Gs
9	$p = CH^2$	$4.000 \cdot 10^{-3}$ Pa $= 0.04$ erg/cm^3	H in Gs

[a] e.m.u. is the electron mass unit ($m_e = 9.108 \cdot 10^{-28}$ g), a.m.u. is the atomic mass unit ($m_a = 1.6605 \cdot 10^{-24}$ g)

Chapter 3
Physics of Atoms and Ions

Abstract General concepts of physics of atoms are considered. A general atom scheme takes the Coulomb interaction of electrons with the nucleus and between them as the strongest interaction which together with the Pauli principle for electrons leads to the shell atom structure. More weak interactions include an exchange interaction of electrons and spin-orbit interaction which determine the scheme of momentum coupling in the atom and its quantum numbers. The detailed analysis is fulfilled for properties of the hydrogen and helium atoms located in the ground and excited states. The analysis of lowest excited states of inert gas atoms is a demonstration of momentum coupling in atoms. Information is collected for spectral parameters of atoms with s and p valence electrons in the ground state.

3.1 Properties of Hydrogen and Helium Atoms and Similar Ions

We first consider the ground and excited states of the hydrogen atom where an electron is located in the Coulomb field of a proton. The electron position in the field of the Coulomb center is described by three space coordinates, namely, r, the distance of the electron from the Coulomb center, θ, the polar electron angle, and φ, the azimuthal angle. In addition, the spin electron state is characterized by the spin electron projection σ onto a given direction. In the non-relativistic limit, space and spin coordinates are separated, and the space wave function has the form

$$\psi(r, \theta, \varphi) = R_{nl}(r)Y_{lm}(\theta, \varphi) \tag{3.1}$$

Here separation of variables is used for an electron located in the Coulomb center field, so that $R_{nl}(r)$ is the radial wave function, and Y_{lm} is the angle wave function. In addition, n is the principal quantum number, l is the angular momentum, m is the angular momentum projection onto a given direction. These quantum numbers are whole values, and $n \geq 1$, $n \geq l+1$, $l \geq 0$, $\mid m \mid \leq l$. The binding state energy ε_n that counts off from the continuous spectrum boundary is given by

© Springer International Publishing AG, part of Springer Nature 2018

B. M. Smirnov, *Atomic Particles and Atom Systems*, Springer Series on Atomic, Optical, and Plasma Physics 51, https://doi.org/10.1007/978-3-319-75405-5_3

$$\varepsilon_n = -\frac{m_e e^4}{2\hbar^2 n^2} \tag{3.2}$$

Here $\hbar$ is the Planck constant, e is the electron charge, m_e is the electron mass, and the nuclear mass assumes to be infinite. As is seen, the states of an electron that is located in the Coulomb field are degenerated with respect to the electron momentum l and its projection m onto a given direction (as well as with respect to the spin projection σ, since the electron Hamiltonian is independent of its spin). For notations of electron states (electron terms) are used its quantum numbers n, l, where the principal quantum number n is given as a value, whereas the quantum number l is denoted by letters, so that notations s, p, d, f, g, h relate to states with the angular momenta $l = 0, 1, 2, 3, 4, 5$ correspondingly. As example, the notation $4f$ respects to a state with quantum numbers $n = 4, l = 3$.

The angle wave function of an electron located in the Coulomb center field is satisfied to the Schrödinger equation

$$\frac{\partial}{\partial \cos \theta} \left(\sin^2 \theta \frac{\partial Y_{lm}}{\partial \cos \theta} \right) + \frac{1}{\sin^2 \theta} \frac{\partial^2 Y_{lm}}{\partial \varphi^2} + l(l+1) Y_{lm} = 0 \tag{3.3}$$

and is given by the formula [44–46]

$$Y_{lm}(\theta, \varphi) = \left[\frac{2l+1}{4\pi} \frac{(l-m)!}{(l+m)!} \right]^{1/2} P_l^m(\cos \theta) \exp(im\varphi) \tag{3.4}$$

These angle wave functions of an electron are represented in Table 3.1.

The radial wave function of an electron in the hydrogen atom is the solution of the Schrödinger equation

$$\frac{1}{r} \frac{d^2}{dr^2} (r R_{nl}) + \left[2\varepsilon + \frac{2}{r} - \frac{l(l+1)}{r^2} \right] R_{nl} = 0, \tag{3.5}$$

that with accounting for boundary conditions has the form

$$R_{nl}(r) = \frac{1}{n^{l+2}} \frac{2}{(2l+1)!} \sqrt{\frac{(n+l)!}{(n-l-1)!}} (2r)^l \exp(-\frac{r}{n}) \ F\left(-n+l+1, \ 2l+2, \ \frac{2r}{n} \right), \tag{3.6}$$

where F is degenerate hypergeometric function. The following normalization condition takes place for the radial wave function

$$\int_0^\infty R_{nl}^2(r) r^2 dr = 1, \tag{3.7}$$

Table 3.2 lists the expressions of the radial electron wave functions for lowest hydrogen atom states.

Table 3.3 contains expressions for average quantities $\langle r^n \rangle$ of the hydrogen atom, where r is the electron distance from the center, and Table 3.4 lists the values of these quantities for lowest states of the hydrogen atom.

$$\langle r^n \rangle = \int_0^\infty R_{nl}^2(r) r^{n+2} dr \qquad (3.8)$$

Fine structure splitting in the hydrogen atom is determined by spin-orbit interaction and corresponds according to its nature to the interaction between spin and magnetic field due to angular electron rotation. The fine structure splitting is given by [44–46]

$$\delta_f = \frac{1}{2} \left(\frac{Z^2 e \hbar}{mc} \right)^2 \left\langle \frac{1}{r^3} \right\rangle \cdot \langle \hat{\mathbf{l}} \hat{\mathbf{s}} \rangle = \frac{2l+1}{4} \left(\frac{Z^2 e \hbar}{mc} \right)^2 \left\langle \frac{1}{r^3} \right\rangle = \left(\frac{e^2}{\hbar c} \right)^2 \frac{Z^4}{2n^3 l(l+1)} \qquad (3.9)$$

Here brackets mean an average over the electron space distribution, and this formula relates to the hydrogenlike ion, where Z is the Coulomb center charge. Since the parameter $[e^2/(\hbar c)]^2/4 = 1.33 \cdot 10^{-5}$ is small, the fine structure splitting is small for the hydrogen atom. Accounting for the fine structure splitting gives the quantum numbers of the hydrogen atom as $lsjm_j$ instead of numbers $lms\sigma$ for degenerate levels, where $s = 1/2$ is the electron spin, $j = l \pm 1/2$ is the total electron momentum, m, σ, m_j are projections of the orbital and total momenta onto a given direction. Note that the lower level corresponds to the total electron momentum $j = l - 1/2$.

The behavior of highly excited or Rydberg states of atoms is close to those of the hydrogen atom, because the coulomb electron interaction with the center is the main interaction. But a short-range electron interaction with an atomic core leads to a displacement of atom energetic levels, and the electron energy instead of that by formula (3.2) is given by [47]

$$\varepsilon_n = -\frac{m_e e^4}{2\hbar^2 n_{ef}^2} = -\frac{m_e e^4}{2\hbar^2 (n - \delta_l)^2}, \qquad (3.10)$$

where n_{ef} is the effective principal quantum number, δ_l is so called quantum defect [48, 49]. Since it is determined by a short-range electron-core interaction, its value decreases with an increase of the electron angular momentum l. Table 3.5 contains the values of the quantum defect for alkali metal atoms [50, 51].

In addition, Fig. 3.1 gives spectrum of the hydrogen atom in the form of the Grotrian diagram.

In order to explain the atom nature, its structure and spectrum, we use one-electron approximation where the atom wave function is the combination of products of one-electron wave functions. In construction the atom wave function we are based on the Pauli exclusion principle [52–54] according to which location of two electrons is prohibited in identical electron states. This means that two electrons with the same spin direction can not be located at the same coordinate. Other formulation of the Pauli exclusion principle is that the total wave function of electrons changes a sign after permutation of two electrons. In this manner the Pauli exclusion principle influences on the atom structure and leads to the exchange interaction potential between electrons. In the case of the helium atom or heliumlike ion one can define the exchange interaction potential as the difference between energies of states with the total spins $S = 0$ and $S = 1$ and identical other quantum numbers.

Let us introduce the exchange interaction potential inside an atom as the difference between states with the total spin $S = 0$ and $S = 1$ for identical other quantum numbers. The Hamiltonian of the helium atom (or heliumlike ions) has the form

$$\hat{H} = -\frac{\hbar^2}{2m_e}\Delta_1 - \frac{\hbar^2}{2m_e}\Delta_2 - \frac{Ze^2}{r_1} - \frac{Ze^2}{r_2} + \frac{e^2}{|\mathbf{r}_1 - \mathbf{r}_2|}, \tag{3.11}$$

where the first two terms describes the kinetic energy of electrons, the third and fourth terms refer to Coulomb interaction of electrons with the nucleus of a charge Z, the last term corresponds to interaction between electrons, and $\mathbf{r}_1$, $\mathbf{r}_2$ are electron coordinates. Correspondingly, the energy E of a given state is

$$E = \langle \Psi \hat{H} | \Psi \rangle, \tag{3.12}$$

where Ψ is thee wave function of electrons for a given state.

Let us construct the wave function of electrons in the one-electron approach in the state with electrons in states $n = 1$ and $n = 2$. The total wave function is a product of the spin wave function and coordinate one. According to the Pauli principle, the total wave function of two electrons changes the sign as a result of permutation of two electrons. Since the spin wave function for the spin $S = 1$ is symmetric with respect to permutation of electrons, the coordinate wave function of this state is antisymmetric with respect to this operation. On contrary, the coordinate wave function for the total spin $S = 0$ is symmetric with respect to permutation of electrons. Let us construct the coordinate wave function for the state, where the principal quantum number $n = 1$ for one electron and $n = 2$ for another electron. The total coordinate wave function which satisfies to the above symmetry with respect permutation of electrons in the one-electron approach has the form

$$\Psi_{0,1}(\mathbf{r}_1, \mathbf{r}_2) = \frac{1}{\sqrt{2}}[\psi(\mathbf{r}_1)\varphi(\mathbf{r}_2) \pm \psi(\mathbf{r}_2)\varphi(\mathbf{r}_1)], \tag{3.13}$$

where $\mathbf{r}_1$, $\mathbf{r}_2$ are coordinates of corresponding electrons, the coordinate wave functions $\psi(\mathbf{r})$ and $\varphi(\mathbf{r})$ correspond to states with $n = 1$ and $n = 2$, sign plus relates to the state with the total spin $S = 0$ and sign minus refers to the state with $S = 1$. From this one can find the difference of energies of these states Δ that is given by

$$\Delta = 2\langle \psi(r_1)\varphi(r_2)]\hat{H} | \psi(r_2)\varphi(r_1) \rangle, \tag{3.14}$$

From this it follows that the total electron spin determines a certain symmetry of the space electron wave function that influence on the electron energy. The exchange interaction (3.14) is determined just by the symmetry of the wave function, rather interactions involving electron spins.

Thus the spectrum of the helium atom (Fig. 3.3) is divided in two independent parts with the total electron spin $S = 0$ and $S = 1$ with an antisymmetric space wave function of electrons in the first case and with symmetric space wave function of electrons in the second case. The radiative transitions between states which belong to different parts are absent practically because of conservation of the total electron spin in radiative transitions due to first approximations in the expansion over a small relativistic parameter. In addition, atom levels related to the total spin $S = 1$ are located below the corresponding levels of states $S = 0$ with identical other quantum numbers. One can see it in the diagram of Fig. 3.3 for the helium atom and in Fig. 3.2 where the lowest excited states of the helium atom are given.

Let us expand the expression (3.11) for the Hamiltonian of electrons by addition to it the exchange interaction and also the spin-orbit interaction. As a result, we represent the electron Hamiltonian in the form

$$\hat{H} = -\frac{\hbar^2}{2m_e}\Delta_1 - \frac{\hbar^2}{2m_e}\Delta_2 - \frac{Ze^2}{r_1} - \frac{Ze^2}{r_2} + \frac{e^2}{|\mathbf{r}_1 - \mathbf{r}_2|} + A\hat{\mathbf{s}}_1\hat{\mathbf{s}}_2 + B\hat{\mathbf{l}}_1\hat{\mathbf{s}}_1 + B\hat{\mathbf{l}}_2\hat{\mathbf{s}}_2 \quad (3.15)$$

As early, an electrostatic interaction of electrons with a core and also with each other, the exchange interaction of electrons, and the interaction of an electron spin with its orbit are included in this Hamiltonian. Spin-orbit interaction is of importance for spectroscopy of excited atom electron states, whereas interaction of a spin with a foreign orbit is not important.

Let us use the electron Hamiltonian (3.15) for the analysis of spectrum of the helium atom that is given in Fig. 3.3 and spectra of heliumlike ions. In the ground atom state both electrons are located in the state $1s$, and the ground atom state is $1s^2$ ^1S, i.e. the total atom spin is zero $S = 0$. The atom state with $S = 1$ and the states $1s$ of both electrons is not realized because the space wave function is antisymmetric with respect to electron permutation and is zero for identical electron states. The quantum number of an excited helium atom is its orbital momentum and spin that can be equal to zero or one. In the latter case the atom quantum number is the total atom momentum that is a sum of the orbital momentum and spin (see Fig. 3.3). In the case of heliumlike ions which consist of the Coulomb center of a charge Z and two electrons, the contribution of different interactions to the total energy varies with variation of the center charge, as it follows from Table 3.6. Correspondingly, the spectrum of heliumlike atoms changes with variation of Z.

3.2   Quantum Numbers of Light Atoms

In considering light atoms, one can ignore relativistic interactions in the first approximation, and the electron Hamiltonian for such an atom in accordance with the expression (3.15) has the form

$$\hat{H} = -\frac{\hbar^2}{2m_e} \sum_i \Delta_i - \sum_i \frac{Ze^2}{r_i} + \sum_{i,k} \frac{e^2}{|\mathbf{r}_i - \mathbf{r}_k|} + A \sum_{i,k} \hat{s}_i \hat{s}_k \qquad (3.16)$$

Here i, k are electron numbers, $\mathbf{r}_i$, $\mathbf{r}_k$ are electron coordinates, $\hat{s}_i$, $\hat{s}_k$ are the operators of electron spins.

The operators of the total electron momentum $\hat{\mathbf{L}}$ and the total electron spin $\hat{\mathbf{S}}$ for an atom are given by

$$\hat{\mathbf{L}} = \sum_i \hat{\mathbf{l}}_i \, , \hat{\mathbf{S}} = \sum_i \hat{\mathbf{s}}_i \, , \qquad (3.17)$$

where $\hat{\mathbf{l}}_i$ and $\hat{\mathbf{s}}_i$ are the operators of the angular momentum and spin for i-th electron. Because these operators commute with the Hamiltonian (3.16) [21], the atom quantum numbers are LSM_LM_S , where L is the angular atom momentum, S is the total atom spin, M_L, M_S are the projections of these momenta onto a given direction.

If we add to the Hamiltonian (3.16) the operator of spin-orbit interaction in the form $B\hat{\mathbf{L}}\hat{\mathbf{S}}$, the energy levels for given atom quantum numbers L and S are split, and the atom eigen states are characterized by the quantum numbers $LSJM_J$, where J is the total atom momentum, M_J is its projection onto a given direction.

The following notations are used to describe an atom state. The number $2S + 1$ of spin projections, the multiplicity of spin states, are given in front above. The angular atom momentum is given by letters S, P, D, F, G etc. for values of the angular momentum $L = 0$, 1, 2, 3, 4, 5 etc. The total atom momentum J is given as a subscript. For example, $^3D_3$ corresponds to atom quantum numbers $S = 1$, $L = 2$, $J = 3$. This method of description of atom states is suitable for light atoms. But this scheme may be used also for heavy atoms to classify atom states, though relativistic effects are of importance for these atoms.

The atom shell model distributes bound electrons over electron shells, so that each shell contains electrons with identical quantum numbers nl. This corresponds to one-electron approximation if the field that acts on a test electron consists of the Coulomb field of the nucleus and a self-consistent field of other electrons. The sequence of shell filling for an atom in the ground state corresponds to that for a center with a screened Coulomb field that acts on a test electron.

We now consider the behavior of energetic parameters for atoms parameters of the lowest electron levels are given for atoms with electron shells p, p^2, p^3, p^4 and p^5 for valence electrons which correspond to atoms of third, fourth, fifth, sixth and seventh groups of the periodic system of elements. This information is represented in Tables 3.7, 3.8, 3.9, 3.10 and 3.11. Table 3.7 contains parameters for atoms of the third group of the periodical system of elements where the ground atom state is $^2P_{1/2}$ in notations of the LS coupling scheme that is accurate for light atoms. The first excitation state refers to the electron term $^2P_{3/2}$ and corresponds to another fine state. Table 3.7 lists the excitation energy ε_{ex} from the ground state to an excited state $^2S_{1/2}$, and also the wavelength $\lambda(^2P_{1/2} - ^2S_{1/2})$ for transition between these states, as well as the radiative lifetime $\tau(^2S_{1/2})$ for an excited state. Note that the fine splitting of levels in the hydrogenlike atom is given by formula (3.9) and depends strongly on the nucleus charge. Table 3.7 contains values of the effective charge Z_{ef} of an internal atom part that determines the fine level splitting.

The statistical weight or the number of states of an atom with different quantum numbers equals $C_6^2 = 15$. in the case of the fourth group of elements with p^2 electron shell, and Fig. 3.4 shows two distributions of electrons of this shell over states with the maximum projections of the atom orbital momentum and the maximum spin projection onto a given direction. As it follows from Fig. 3.4, one of electron terms in this case which refers to the maximum orbital momentum, is characterized by the orbital momentum projection onto a given direction $M_L = 2$. This electron term is ^1D, and the second electron term that corresponds to the maximum spin is ^3P. The total number of states related to these electron terms equals 14, and one more electron term is ^1S. The ground electron state of atoms of this group is $^3P_0$, and Table 3.8 contains both the energies of excitation of other electron terms (^1D and ^1S), and the energy of excitation of other fine states ($^3P_1$ and $^3P_2$). Evidently, in all the cases the energies of excitation of fine states is less than that for excitation other electron terms of a given electron shell.

Two distributions of electrons of the electron shell p^3 (fifth group of the periodical system of elements) over states are shown in Fig. 3.5, so that the first state is characterized by the maximum projection of the total orbital momentum of electrons onto a given direction, and the maximum projection of the total electron spin corresponds to the second distribution. In the case of the fifth group of elements with electron shell Then the state with the maximum orbital momentum projection relates to the electron term ^2D, and the state with the maximum spin projection refers to the electron term ^4S. The total number of electron states for fifth group of the periodical system of elements with the electron shell p^3 is equal $C_6^3 = 20$. Evidently, one more electron term whose statistical weight equals 6, is ^2P. The ground state of this group atoms is $^4S_{3/2}$. Table 3.9 contains the excitation energies of electron states of atoms with the electron shell p^3 and also fine splitting of excited levels.

One can construct the sequence of atom levels for sixth group of the periodical system of elements with the electron shell p^4 on the basis of those for the fourth group by replacing of $p-$ electrons with $p-$ holes in the completed electron shell p^6. In this case we obtain the same sequence of electron terms, but the opposite sequence of fine states. Hence, the ground electron state for atoms of the sixth group of the periodical system of elements is $^3P_2$. Table 3.10 gives the excitation energies of atoms of the sixth group, as well as the fine splitting of levels for the lowest electron term.

Atoms of halogens which relate to the seventh group of the periodical system of elements with the electron shell p^5 are characterized by electron terms which are analogous to atoms with one p-valence electron because they contains one p-hole with respect to the completed electron shell. Hence, the electron term of the ground state of halogen atoms coincides with that of atoms of the third group of the periodic system of elements, but sublevels of fine structure have the inverse sequence. The ground state of halogen atoms is $^2P_{3/2}$. Table 3.11 exhibits parameters of lowest states of halogen atoms.

One can also represent information about energetic parameters for the lowest states of some atoms with non-completed electron shells p^3, p^4 and p^5 on diagrams. The diagrams of Figs. 3.6, 3.7, 3.8, 3.9, 3.10, 3.11, 3.12, 3.13 give the information of such a type and allows one to check the above scheme. In particular, fine splitting of the ground state of the tellurium atom (Fig. 3.13) is approximately more in 20 times than that for the oxygen atom with the same electron structure (Fig. 3.8) and the character of fine splitting in these cases is different. Nevertheless using the same classification method for both cases is convenient.

Valence electrons determine properties of atoms and atom interaction. One can construct the valence electron shell to be consisted of one electron and atomic rest. This parentage scheme allows one to analyze atom properties due to one-electron transitions. Then the electron wave function of an atom $\Psi_{LSM_LM_S}(1, 2, \ldots, n)$ is expressed through the wave function of an atomic rest $\Psi_{L'S'M'_LM'_S}(2, \ldots, n)$ and the wave function of a valence electron $\psi_{l\frac{1}{2}m\sigma}(1)$ in the following manner [55–57]

$$\Psi_{LSM_LM_S}(1, 2, \ldots, n) = \frac{1}{\sqrt{n}}\hat{P} \sum_{L'M'_LS'M'_Sm\sigma} G^{LS}_{L'S'}(l, n)$$

$$\begin{bmatrix} l & L' & L \\ m & M'_L & M_L \end{bmatrix}\begin{bmatrix} \frac{1}{2} & S' & S \\ \sigma & M'_S & M_S \end{bmatrix}\psi_{l\frac{1}{2}m\sigma}(1) \cdot \Psi_{L'S'M'_LM'_S}(2, \ldots, n),$$

Here n is a number of valence electrons, the operator $\hat{P}$ permutes coordinates and spin of a test electron that is denoted by an argument 1 with those of valence electrons of the atomic rest; LSM_LM_S are the atom quantum numbers, $L'S'M'_LM'_S$ are quantum numbers of the atomic rest, $l\frac{1}{2}m\sigma$ are quantum numbers of a test electron, $G^{LS}_{L'S'}(l, n)$ is the parentage or Racah coefficient [55], and the Clebsh-Gordan coefficients result from summation of momenta (orbital and spin momenta) of a test electron and atomic rest into an atom momentum.

Note that the one-electron wave function by analogy with formula (3.1) has the form

$$\psi(r, \theta, \varphi) = R_{nl}(r)Y_{lm}(\theta, \varphi)\chi_{1/2,\sigma} \qquad (3.18)$$

where the angular wave function Y_{lm} is given by formula (3.4), $\chi_{1/2,\sigma}$ is the spin electron wave function, and the radial wave function $R_{nl}(r)$ is normalized by the condition (3.6) and satisfies to the Schrödinger equation (3.5) far from the core. The Coulomb interaction potential between the electron and core takes place at large electron distances r from the core where the solution of this equation has the form

$$R_{nl} = Ar^{\frac{1}{\gamma}-1}\exp(-\gamma r), \quad r\gamma \gg 1; \quad r\gamma^2 \gg 1 \tag{3.19}$$

Here $\gamma = \sqrt{(-2\varepsilon)}$ in atomic units, and the electron behavior far from the core is determined by two asymptotic parameters, an exponent γ and an amplitude A.

The parentage scheme relates to a non-relativistic approximation when the orbital momentum and spin are quantum numbers for an atom and its atomic rest, and this scheme accounts for the spherical atom symmetry if non-relativistic interactions are small. Then $LSM_L M_S$, i.e. the atom angular momentum, spin, and their projections on a given direction are the quantum numbers. In this approximation the atom states are degenerated over the angular momentum and spin projections, i.e. identical energies corresponds to different projections on a given direction for these momenta. The simplest construction of electron shells relates to s and $p-$ valence electrons, and the values of parentage coefficients for these cases are given in Table 3.12.

Fractional parentage coefficients satisfy to some relations that follow from the definition of these quantities. The normalization of the atomic wave function takes the form

$$\sum_{L'S'v} \left[G_{L'S'}^{LS}(l, n, v) \right]^2 = 1 \tag{3.20}$$

The total number of valence electrons equals to $4l + 2$ if the shell is completed. The analogy between electrons and vacancies connects the parameters of electron shells in the following method [55, 57]

$$G_{L'S'}^{LS}(l, n, v) = (-1)^{L+L'+S+S'-l-1/2} \cdot \left[\frac{n(2S'+1)(2L'+1)}{(4l+3-n)(2S+1)(2L+1)} \right]^{1/2} G_{L'S'}^{LS}(l, 4l+3-n, v) \tag{3.21}$$

The parentage scheme relates to light atoms where relativistic effects are small. Then within the framework of LS-scheme of momentum summation the angular atom momentum is the sum of angular momenta of an atomic rest and test electron, and the atom spin is the sum of spins of these particles. Then as a result of interaction of an atom angular momentum L and spin S, these momenta are summed into the total atom momentum J, so that atom quantum numbers are $LSJM_J$, where M_J is the projection of the total atom momentum on a given direction. As is seen, splitting of the electron terms with given values of L and S due to spin-orbit interaction leads to atom quantum numbers LSJ in contrast to quantum numbers LS in neglecting spin-orbit interaction.

Table 3.13 contains the number of electron terms, the number of electron levels and the statistical weight of atoms with non-completed electron shells. Here the statistical weight g, i.e. the total number of electron states, for a non-completed electron shell contained k shells with the angular momentum l is equal

$$g = C_{4l+2}^k \tag{3.22}$$

This is a number of ways to distribute the electrons over states. We call the number of electron terms in Table 3.13 as the number of degenerate states in neglecting the spin-orbit interaction, and the number of levels takes into account fine splitting of levels. For example, if a non-completed electron shell contains one d-electrons, one level corresponds to the electron term ^2D, and there are two electron terms with accounting for the fine splitting, namely, $^2D_{3/2}$ and $^2D_{5/2}$. For definiteness, Table 3.14 contains the number of electron states and represent electron terms for atoms with valence d-electrons in neglecting the spin-orbit interaction in the case of non-completed d-electron shells. In particular, electron shells d^2 and d^8 which relate to the seventh line of Table 3.13, are characterized by 9 electron terms, namely, $^1S_0, \, ^3P_0, \, ^3P_1, \, ^3P_2,$ $^1D_2, \, ^3F_2, \, ^3F_3, \, ^3F_4, \, ^1G_4$. A number of states in this case is the sum of momentum projections, that is $1 + 1 + 3 + 5 + 5 + 5 + 7 + 9 + 9$ in this case for indicated electron terms.

The parentage scheme is simple for valence s and p-electrons, and the values of fractional parentage coefficients are represented in Table 3.12 for these cases. In the case of d and f valence electrons, removal of one electron can lead to different states of an atomic rest at identical values of the atom angular momentum and spin. To distinguish these states, one more quantum number v, the seniority, is introduced.

The ground atom state for a given electron shell follows from the empiric Hund law [58, 59] according to which the maximum atom spin corresponds to the ground electron state, and the total atom momentum is minimal if the shell is filled below one half and is maximal from possible ones if the electron shells is filled more than by half. For example, the nitrogen atom with the electron shell p^3 has three electron terms, $^4S, \, ^2D, \, ^2P$ and the electron term ^4S corresponds to its ground state. The ground state of the aluminium atom with the electron shell p is characterized by the total momentum $1/2$ (the state $^2P_{1/2}$), and the ground state of the chlorine atom with the electron shell p^5 is $^2P_{3/2}$ (the total momentum is $3/2$). One can convince in the validity of the Hund law for the ground atom state for atoms with filling electron $p-$shell on the basis of data of Tables 3.13, 3.14 and 3.15. The ground states of atoms with d and f filling electron shells are given in Tables 3.15 and 3.16.

Note that the parentage scheme holds true for light atoms both for the ground and lower excited atom states, and it is violated for heavy atoms both due to an increase of the role of relativistic effects and because of overlapping of electron shells with different nl (Fig. 3.14).

In construction of electron shells we were based on the LS-scheme of electron momentum summation. Along with this, jj-scheme of electron momentum summation is possible. In particular, taking an atom to be composed from an atomic rest with the angular momentum $\mathbf{l}$ and spin $\mathbf{s}$ and from a valence or excited electron with the angular momentum $\mathbf{l}_e$ and spin $\mathbf{s}_e$, one can realize the summation of these momenta into the total atom momentum in these schemes

$$\mathbf{l} + \mathbf{l}_e = \mathbf{L}, \quad \mathbf{s} + \mathbf{s}_e = \mathbf{S}, \, \mathbf{S} + \mathbf{L} = \mathbf{J} \quad (LS\text{-coupling})$$
$$\mathbf{l} + \mathbf{s} = \mathbf{j}, \quad \mathbf{l}_e + \mathbf{s}_e = \mathbf{j}_e \, \mathbf{j} + \mathbf{j}_e = \mathbf{J} \quad (jj\text{-coupling})$$

A certain hierarchy of interaction corresponds to each scheme of momentum summation, and we will follow this connection. Let us represent the Hamiltonian of interacting electron and atomic rest in the form

$$\hat{H} = \hat{H}_{core} + \hat{H}_e - a\hat{\mathbf{l}}\hat{\mathbf{s}} - b\hat{\mathbf{s}}\hat{s}_e + c\hat{\mathbf{l}}_e\hat{s}_e \tag{3.23}$$

The terms $\hat{H}_{core}$ and $\hat{H}_e$ in this sum accounts for the electron kinetic energy and electrostatic interaction for an atomic rest and valence electron correspondingly, the forth term describes an exchange interaction of a valence electron and atomic rest, and third and fifth terms are responsible for spin-orbit interaction for an atomic rest and valence electron. It is of importance that this Hamiltonian commutes with the total atom momentum [60]

$$\hat{\mathbf{J}} = \hat{\mathbf{l}} + \hat{\mathbf{s}} + \hat{\mathbf{l}}_e + \hat{s}_e \tag{3.24}$$

This means that the total atom momentum is the quantum atom number for both schemes of momentum summation at any relation between exchange and spin-orbit interactions. Next, the LS coupling scheme is valid in the limit of a weak spin-orbit interaction, whereas the jj coupling scheme corresponds to a weak exchange interaction. The LS-coupling scheme considered above is valid for light atoms, where relativistic interactions are small, and the jj coupling scheme may be used as a model for heavy atoms. Nevertheless, it is used often for notations of electron terms of heavy atoms, and Table 3.17 contains electron terms of atoms with filling p shells for both coupling schemes. Notations for jj coupling scheme are such that the total momentum of an individual electron is indicated in square parentheses and the total number of such electrons is given as a right superscript.

Thus, roughly in accordance with the hierarchy of interaction inside the atom we have two basic types of momentum coupling. If the electrostatic interaction exceeds a typical relativistic interaction, we have the LS-type of momentum summation, in another case we obtain the jj type of momentum coupling. Really, the electrostatic interaction V_{el} inside the atom corresponds to splitting of atom levels of the same electron shell, while the relativistic interaction may be characterized by the fine splitting δ of atom levels.

The specific case relates to atoms of inert gases, and Fig. 3.12 gives the lowest group of excites states of the argon atom, and one can demonstrate on this example the ways of classifications of these states. The structure of the electron shell for these states is $3p^5 4s$, and along with notations of the LS coupling scheme the notations of the jj coupling scheme are used. In the last case the total momentum of the atomic core is given inside square parentheses, and the total atom momentum is given as a right subscript. In addition, the Pashen notations are used for lowest excited states of inert gas atoms. Then the electron terms of the lowest group of excited states are denoted as $1s_5$, $1s_4$, $1s_3$, $1s_2$, and the subscript decreases with an increase of excitation.

We also consider the lowest excited states of inert gases which are given in Figs. 3.15, 3.16, 3.17 and 3.18 with Pashen notations for the next group of excited states with the electron shell $np^5(n+1)p$ (n is the principal quantum number of electron of the valence shell) are from $2p_{10}$ up $2p_1$ as the excitation energy increases. In notations of the jj scheme of momentum coupling, the state of an excited electron is given before the core state.

3.3   Lowest Excited States of Inert Gas Atoms

As an example, we consider the lowest excited states of inert gas atoms with the electron shell $np^5 (n+1)s$. Because the momentum of a valence electron is zero, one can represent the interaction operator

$$\Delta \hat{H} = -a\hat{\mathbf{l}}\hat{\mathbf{s}} - b\hat{\mathbf{s}}\hat{\mathbf{s}}_e, \tag{3.25}$$

where **l** is the operator of the core orbital momentum, **s** is the operator of core spin, and $\mathbf{s}_e$ is the operator of the valence electron spin momentum. The first term corresponds to spin-orbit interaction, and the second term describes exchange interaction between a core and valence electron due to symmetry of the wave function. In the case $b \gg a$, the LS-coupling scheme holds true, while in the case $a \gg b$ the jj-coupling scheme is valid. It is considered a general case where the above interactions are compared.

In this scheme, where we account for only spin-orbit and exchange interactions, positions of energy levels on the basis of the Schrödinger equation

$$\Delta \hat{H} \Psi = \varepsilon \Psi \tag{3.26}$$

Counting out the level positions from the lowest level of this group, i.e. taking $\varepsilon_5 = 0$, one can find for the other three levels [61]

$$\varepsilon_{2,4} = \frac{3}{4}a + \frac{1}{2}b \pm \frac{1}{4}\sqrt{9a^2 - 4ab + 4b^2}, \;\; \varepsilon_3 = \frac{3}{2}a \tag{3.27}$$

One can use these expressions for comparison with real positions [8, 15, 62] of the corresponding levels. This allows us to determine the values of parameters of this scheme (3.25) and also to estimate its accuracy. Indeed, (3.27) give three relations for determination of parameters a and b. In addition, the parameter a follows from the fine splitting of ion levels Δ_f which coincides with the excitation energy $\varepsilon_3 - \varepsilon_5$ of the state $1s_3$. Comparison of these values is given in Table 3.18, and if they coincide, the parameter $\delta = (\varepsilon_3 - \varepsilon_5)/\Delta_f$ is equal to one. Next, one can determine the parameter b from (3.27)

$$b = \varepsilon_2 + \varepsilon_4 - \varepsilon_3 - \varepsilon_5 \tag{3.28}$$

One can see that if the scheme (3.25) is valid, the quantities $\varepsilon_2 - \varepsilon_4$, $C = \sqrt{9a^2/4 - ab + b^2}$ and $D = \sqrt{\Delta_f^2 - 2\Delta_f b/3 + b^2}$ are identical. The ratios of these values are given in Table 3.18.

Note that LS-coupling scheme is valid in the case $b \gg a$, and the jj-coupling of momenta holds true under condition $a \gg b$, and for lower excited states of inert gas atoms an intermediate case takes place within the framework of the scheme (3.25). In addition, this scheme is violated for heavy atoms, since, on the one hand, new relativistic interactions are essential, and, on the other hand, the energies of excitation of next states become comparable with the width of this energy band. In particular, as it follows from Fig. 3.18, the energy difference for states $1s_2$ and $1s_5$ is roughly

$10^4 \, cm^{-1}$, where the transition energy between $1s_5$ and $2p_{10}$ state is approximately $10^2 \, cm^{-1}$. This exhibits a strong interaction between states of different groups.

3.4 Parameters of Atoms and Ions in the Form of Periodical Tables

It is convenient to give information for atoms or atomic system of various elements in the form of periodical tables. This allows one, on the one hand, to insert large information in a simple scheme, and, on the other hand, to have a simple scheme in order to read this information. The above analysis of the hydrogen and helium atoms allows one to ascertain the character of influence of interaction between valence electrons charged nuclei that includes the Coulomb and exchange interactions. Diagram of Figs. 3.19 and 3.20 give the ionization potentials for atoms and their first ions in the ground states on the basis of the periodical table of elements together with construction of the electron shell and the electron term for the ground atom state within the framework of the LS-scheme of momentum coupling. Because of the periodical character of the electron shell structure, we obtain the periodical dependence for atom parameters [63, 64]. Figure 3.21 contains values of the electron affinity of atoms, i.e. the binding energy of a valence electron in the negative ion. Because of a short-range interaction for a valence electron and an atomic core, in contrast to atoms, a number of bound states for a negative ion is finite (it is often one or zero). The binding energies of excited states of negative ions are also represented in diagram of Fig. 3.21 in the cases when they exist (in particular, for elements of the fourth group). Diagram of Fig. 3.22 gives parameters of lower excited states of atoms on the basis of the periodical table, and the splitting energies for lower atom states are contained in diagram of Fig. 3.23.

On the basis of the above analysis one can understand the atom construction. An atom is a system of electrons located in the Coulomb field of a charged nucleus with accounting for the Pauli exclusion principle. In the first approach, electrons are distributed in the nucleus field over electron shells. The exchange interaction of electrons that is determined by the symmetry of the electron wave function and the spin-orbit interaction becomes significantly in the second approach. As a result along the electron distribution over electron shells, quantum numbers of valence electrons are JM_J, i.e. the total electron moment and its projection onto a given direction. This holds true for a not heavy atoms until other relativistic interactions, except spin-orbit interaction, become essential. For light atoms their quantum numbers are also LS, i.e. the orbital atom momentum and its spin. Note also that because the total number of electron states is determined by valence electrons on a non-completed electron shell, notations within the framework of the LS scheme of momentum summation are suitable in the case where this scheme does not hold true.

Figures

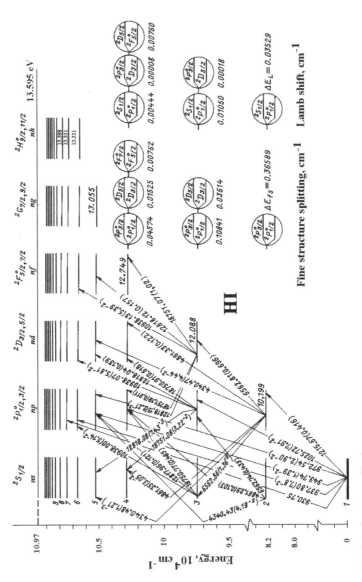

Fig. 3.1 Spectrum of hydrogen atom

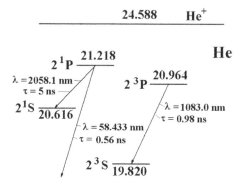

Fig. 3.2 Lowest excited states of the helium atom. The excitation energy for a given state is expressed in eV, the wavelengths λ and radiative lifetimes τ are indicated for a corresponding radiative transition

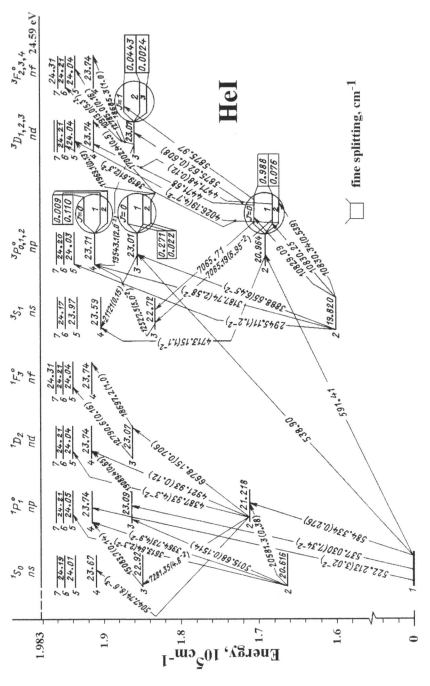

Fig. 3.3 Spectrum of helium atom

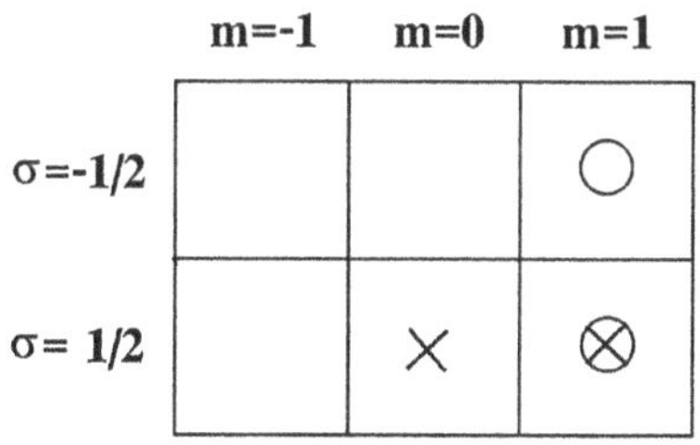

	m=-1	m=0	m=1
σ=-1/2			○
σ= 1/2		✕	⊗

Fig. 3.4 Distribution of valence p-electrons over p^2 electron shell. σ is the spin projection of a given electron onto a given axis, and m is the orbital momentum projection of this electron onto the axis. Two electrons are distributed over possible states in accordance with the Pauli exclusion principle

	m=-1	m=0	m=1
σ=-1/2			○
σ= 1/2	✕	⊗	⊗

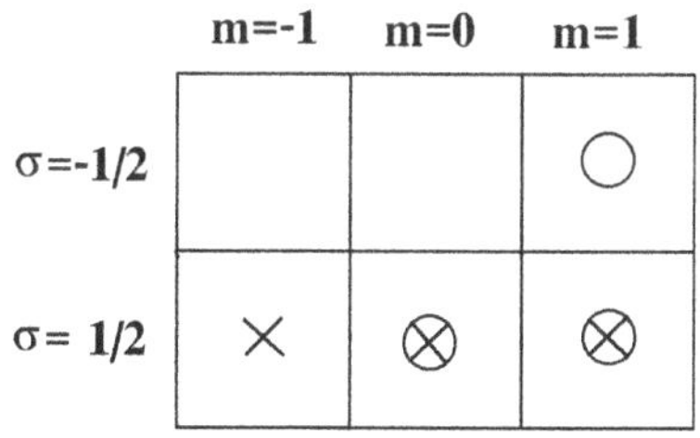

Fig. 3.5 Distribution of valence p-electrons over p^3 electron shell; σ is the spin projection of a given electron onto a given axis, and m is the orbital momentum projection of this electron onto the axis. Three electrons are distributed over possible states in accordance with the Pauli exclusion principle

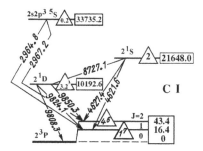

Fig. 3.6 The lowest excited states of the carbon atom. Energies of excitation for corresponding states are indicated inside rectangle boxes with accounting for their fine structure and are expressed in cm^{-1}, wavelengths of radiative transitions are given inside arrows and are expressed in Å, the radiative lifetimes of states are placed in triangle boxes and are expressed in s, and 1^7 means $1 \cdot 10^7$

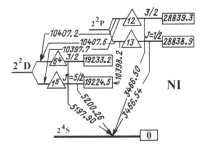

Fig. 3.7 The lowest excited states of the nitrogen atom. Energies of excitation for corresponding states are indicated inside rectangle boxes with accounting for their fine structure and are expressed in cm^{-1}, wavelengths of radiative transitions are given inside arrows and are expressed in Å, the radiative lifetimes of states are placed in triangle boxes and are expressed in s, 1^5 means $1 \cdot 10^5$

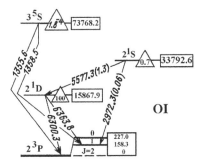

Fig. 3.8 The lowest excited states of the oxygen atom. Notations are similar to those of Fig. 3.7, the oscillator strength of a radiative transition is given in parentheses near the wavelength

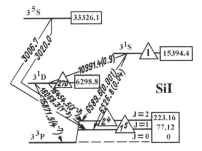

Fig. 3.9 The lowest excited states of the silicon atom. Notations are similar to those of Figs. 3.7 and 3.8

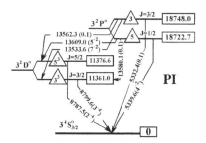

Fig. 3.10 The lowest excited states of the phosphorus atom. Notations are similar to those of Figs. 3.7 and 3.8

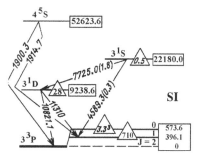

Fig. 3.11 The lowest excited states of the sulfur atom. Notations are similar to those of Figs. 3.7 and 3.8

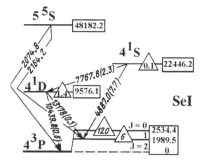

Fig. 3.12 The lowest excited states of the selenium atom. Notations are similar to those of Figs. 3.7 and 3.8

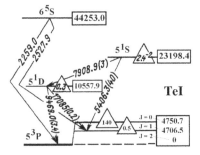

Fig. 3.13 The lowest excited states of the tellurium atom. Notations are similar to those of Figs. 3.7 and 3.8

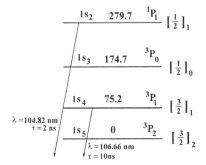

Fig. 3.14 The lowest excited states of the argon atom. Excitation energies count off from the lowest excited state and are given in cm^{-1}, the wavelengths λ and radiative lifetimes τ are indicated for a corresponding radiative transition

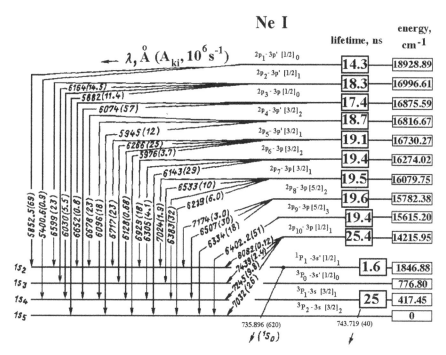

Fig. 3.15 The lowest excited states of the neon atom. Energies of excitation for corresponding states are indicated inside right rectangle boxes and are expressed in cm^{-1}, radiative lifetimes of corresponding states are given inside left rectangle boxes in *ns*, wavelengths of radiative transitions are indicated inside arrows and are expressed in Å, the rates of radiative transitions re represented in parentheses near the wavelength and are expressed in 10^6 s^{-1}

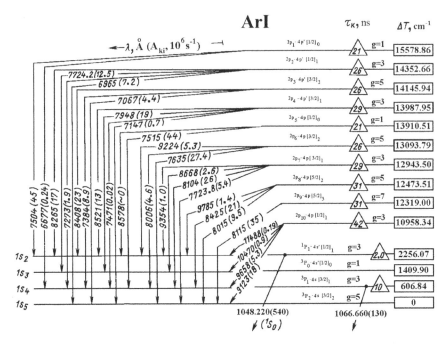

Fig. 3.16 The lowest excited states of the argon atom. Notations are similar to those of Fig. 3.15

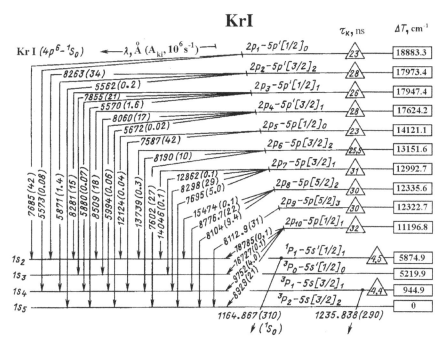

Fig. 3.17 The lowest excited states of the krypton atom. Notations are similar to those of Fig. 3.15

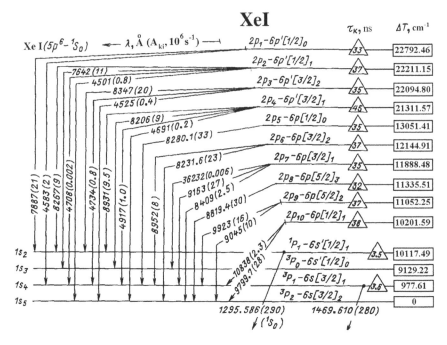

Fig. 3.18   The lowest excited states of the xenon atom. The notations are similar to those of Fig. 3.15. Three types of notations are used, namely, Pashen notations (left), notations for *LS*-coupling (right with respect to values), and notations for *jj*- coupling (right)

Ionization potentials of

Group / Period	I	II	III	IV	V
1	1.008 1s 13.598 $_1$H $^2S_{1/2}$ Hydrogen	24.588 1s² 4.003 / 54.418 $^1S_0$ $_2$He Helium			
2	6.491 2s 5.392 / 75.64 / 122.4 $_3$Li $^2S_{1/2}$ Lithium	9.012 2s² 9.323 / 18.211 / 153.9 / 217.7 $_4$Be $^1S_0$ Berillium	8.298 2p 10.81 / 25.155 / 37.931 / 259.38 $^2P_{1/2}$ $_5$B Boron	11.260 2p² 12.011 / 24.384 / 47.89 / 64.49 $^3P_0$ $_6$C Carbon	14.534 2p³ 14.007 / 29.602 / 47.45 / 77.47 $^4S_{3/2}$ $_7$N Nitrogen
3	22.990 3s 5.139 / 47.287 / 71.620 / 98.92 $_{11}$Na $^2S_{1/2}$ Sodium	24.305 3s² 7.646 / 15.035 / 80.144 / 109.27 $_{12}$Mg $^1S_0$ Magnesium	5.986 3p 26.982 / 18.829 / 28.448 / 119.99 $^2P_{1/2}$ $_{13}$Al Aluminium	8.152 3p² 28.086 / 16.346 / 33.493 / 45.142 $^3P_0$ $_{14}$Si Silicon	10.487 3p³ 30.974 / 19.770 / 30.203 / 51.444 $^4S_{3/2}$ $_{15}$P Phosphorus
4	39.098 4s 4.341 / 31.63 / 45.81 / 60.91 $_{19}$K $^2S_{1/2}$ Potassium	40.08 4s² 6.113 / 11.872 / 50.913 / 67.3 $_{20}$Ca $^1S_0$ Calsium	44.956 3d4s² 6.562 / 12.800 / 24.757 / 73.49 $_{21}$Sc $^2D_{3/2}$ Scandium	47.88 3d²4s² 6.82 / 13.58 / 27.49 / 43.27 $_{22}$Ti $^3F_2$ Titanium	50.942 3d³4s² 6.74 / 14.66 / 29.31 / 46.71 $_{23}$V $^4F_{3/2}$ Vanadium
(4 cont.)	7.726 3d¹⁰4s 63.546 / 20.293 / 36.84 / 57.4 $^2S_{1/2}$ $_{29}$Cu Copper	9.394 4s² 65.38 / 17.964 / 39.72 / 59.57 $^1S_0$ $_{30}$Zn Zinc	5.999 4p 69.72 / 20.51 / 30.7 / 64.2 $^2P_{1/2}$ $_{31}$Ga Gallium	7.900 4p² 72.59 / 15.935 / 34.2 / 45.7 $^3P_0$ $_{32}$Ge Germanium	9.789 4p³ 74.922 / 18.59 / 28.4 / 50.1 $^4S_{3/2}$ $_{33}$As Arsenic
5	85.468 5s 4.177 / 27.290 / 39.2 / 52.6 $_{37}$Rb $^2S_{1/2}$ Rubidium	87.62 5s² 5.694 / 11.030 / 42.88 / 56.28 $_{38}$Sr $^1S_0$ Strontium	88.906 4d5s² 6.217 / 12.24 / 20.525 / 60.61 $_{39}$Y $^2D_{3/2}$ Yttrium	91.22 4d²5s² 6.837 / 13.13 / 23.1 / 34.419 $_{40}$Zr $^3F_2$ Zirconium	92.906 4d⁴5s 6.88 / 14.32 / 25.0 / 37.7 $_{41}$Nb $^6D_{1/2}$ Niobium
(5 cont.)	7.576 4d¹⁰5s 107.87 / 21.49 / 34.8 $^2S_{1/2}$ $_{47}$Ag Silver	8.994 5s² 112.41 / 16.908 / 37.47 $^1S_0$ $_{48}$Cd Cadmium	5.786 5p 114.82 / 18.87 / 28.0 / 57.0 $^2P_{1/2}$ $_{49}$In Indium	7.344 5p² 118.69 / 14.632 / 30.50 / 40.74 $^3P_0$ $_{50}$Sn Tin	8.609 5p³ 121.75 / 16.53 / 25.32 / 44.16 $^4S_{3/2}$ $_{51}$Sb Antimony
6	132.90 6s 3.894 / 23.15 / 33.4 / 46 $_{55}$Cs $^2S_{1/2}$ Cesium	137.33 6s² 5.212 / 10.004 / 35.8 / 47 $_{56}$Ba $^1S_0$ Barium	138.90 5d6s² 5.577 / 11.1 / 19.18 / 49.9 $_{57}$La $^2D_{3/2}$ Lanthanum	178.49 5d²6s² 6.8 / 14.9 / 23.3 / 33.4 $_{72}$Hf $^3F_2$ Hafnium	180.95 5d³6s² 7.89 $_{73}$Ta $^4F_{3/2}$ Tantalum
(6 cont.)	9.226 5d¹⁰6s 196.97 / 20.5 / 34 / 43 $^2S_{1/2}$ $_{79}$Au Gold	10.438 5d¹⁰6s² 200.59 / 18.76 / 34.2 / 46 $^1S_0$ $_{80}$Hg Mercury	6.108 6p 204.38 / 20.43 / 29.85 $^2P_{1/2}$ $_{81}$Tl Thallium	7.417 6p² 207.2 / 15.033 / 31.94 / 42.33 $^3P_0$ $_{82}$Pb Lead	7.286 6p³ 208.98 / 16.7 / 25.56 / 45.3 $^4S_{3/2}$ $_{83}$Bi Bismuth
7	[223] 7s 4.0 $_{87}$Fr $^2S_{1/2}$ Francium	226.02 7s² 5.278 / 10.15 $_{88}$Ra $^1S_0$ Radium	227.03 6d7s² 5.2 / 11.75 / 20 $_{89}$Ac $^2D_{3/2}$ Actinium		

Fig. 3.19 Ionization potentials of atoms and their ions

atoms and their ions

Legend (Shell of valence electrons):

Atomic weight — 95.94 | 4d⁵5s — 7.099
Symbol — Mo | 7S₃ — 16.16 / 27.2
Atomic number — 42 | — 46.4
Molibdenum
Element   Electron term

Ionization potentials of the atom and first three ions in eV

VI	VII	VIII
13.618 / 35.118 / 54.936 / 77.414 — 2p⁴ 15.999 — ³P₂ ₈O — Oxygen	17.423 / 34.971 / 62.71 / 87.14 — 2p⁵ 18.998 — ²P₃/₂ ₉F — Fluorine	21.565 / 40.963 / 63.46 / 97.12 — 2p⁶ 20.179 — ¹S₀ ₁₀Ne — Neon
10.360 / 23.338 / 34.83 / 47.305 — 3p⁴ 32.06 — ³P₂ ₁₆S — Sulfur	12.968 / 23.814 / 39.61 / 53.47 — 3p⁵ 35.453 — ²P₃/₂ ₁₇Cl — Chlorine	15.760 / 27.630 / 40.911 / 59.81 — 2p⁶ 39.948 — ¹S₀ ₁₈Ar — Argon
51.996 3d⁵4s 6.766 / 16.50 / 31.0 / 49.2 — ₂₄Cr ⁷S₃ — Chromium	54.938 3d⁵4s² 7.434 / 15.640 / 33.67 / 51.2 — ₂₅Mn ⁶S₅/₂ — Manganese	55.847 3d⁶4s² 7.902 / 16.188 / 30.65 / 54.8 — ₂₆Fe ⁵D₄ — Iron	58.933 3d⁷4s² 7.86 / 17.084 / 33.5 / 51.3 — ₂₇Co ⁴F₉/₂ — Cobalt	58.69 3d⁸4s² 7.637 / 18.169 / 35.3 / 54.9 — ₂₈Ni ³F₄ — Nickel
9.752 / 21.16 / 30.82 / 42.95 — 4p⁴ 78.96 — ³P₂ ₃₄Se — Selenium	11.814 / 21.81 / 35.90 / 47.3 — 4p⁵ 79.904 — ²P₃/₂ ₃₅Br — Bromine	14.000 / 24.360 / 36.95 / 52.5 — 4p⁶ 83.80 — ¹S₀ ₃₆Kr — Krypton
95.94 4d⁵5s 7.099 / 16.16 / 27.2 / 46.4 — ₄₂Mo ⁷S₃ — Molibdenum	[98] 4d⁵5s² 7.28 / 15.26 / 29.5 — ₄₃Tc ⁶S₅/₂ — Technetium	101.07 4d⁷5s 7.366 / 16.76 / 28.5 — ₄₄Ru ⁵F₅ — Ruthenium	102.91 4d⁸5s 7.46 / 18.08 / 31.1 — ₄₅Rh ⁴F₉/₂ — Rhodium	106.42 4d¹⁰ 8.336 / 19.43 / 32.9 — ₄₆Pd ¹S₀ — Palladium
9.010 / 18.6 / 27.96 / 37.42 — 5p⁴ 127.60 — ³P₂ ₅₂Te — Tellurium	10.451 / 19.131 / 33.0 — 5p⁵ 126.90 — ²P₃/₂ ₅₃I — Iodine	12.130 / 20.98 / 31.0 / 45 — 5p⁶ 131.29 — ¹S₀ ₅₄Xe — Xenon
183.85 5d⁴6s² 7.98 — ₇₄W ⁵D₀ — Tungsten	186.21 5d⁵6s² 7.88 — ₇₅Re ⁶S₅/₂ — Rhenium	190.2 5d⁶6s² 8.73 — ₇₆Os ⁵D₄ — Osmium	192.22 5d⁷6s² 9.05 — ₇₇Ir ⁴F₉/₂ — Iridium	195.08 5d⁹6s 8.96 / 18.56 — ₇₈Pt ³D₃ — Platinum
8.417 — 6p⁴ [209] — ³P₂ ₈₄Po — Polonium	9.0 — 6p⁵ [210] — ²P₃/₂ ₈₅At — Astatine	10.75 — 6p⁶ [222] — ¹S₀ ₈₆Rn — Radon

Fig. 3.19 (continued)

Lanthanides

Config		Mass		Z · Symbol · Term			Name
$4f5d6s^2$	5.539	140.12	10.8	$_{58}Ce\ ^1G_4$	20.20	39.76	Cerium
$4f^36s^2$	5.47	140.91	10.6	$_{59}Pr\ ^4I_{9/2}$	21.62	38.98	Praseodymium
$4f^46s^2$	5.525	144.24	10.7	$_{60}Nd\ ^5I_4$	22.1	40.4	Neodymium
$4f^56s^2$	5.58	[145]	10.9	$_{61}Pm\ ^6H_{5/2}$	22.3	41.0	Promethium
$4f^66s^2$	5.644	150.36	11.1	$_{62}Sm\ ^7F_0$	23.4	41.4	Samarium
$4f^76s^2$	5.670	151.96	11.24	$_{63}Eu\ ^8S_{7/2}$	24.9	42.7	Europium
$4f^75d6s^2$	6.150	157.25	12.1	$_{64}Gd\ ^9D_2$	20.6	44.0	Gadolinium
$4f^96s^2$	5.864	158.92	11.5	$_{65}Tb\ ^6H_{15/2}$	21.9	39.4	Terbium
$4f^{10}6s^2$	5.939	162.50	11.7	$_{66}Dy\ ^5I_8$	22.8	41.4	Dysprosium
$4f^{11}6s^2$	6.022	164.93	11.8	$_{67}Ho\ ^5I_{15/2}$	22.8	42.5	Holmium
$4f^{12}6s^2$	6.108	167.26	11.9	$_{68}Er\ ^3H_6$	22.7	42.7	Erbium
$4f^{13}6s^2$	6.184	168.93	12.1	$_{69}Tm\ ^2F_{7/2}$	23.7	42.7	Thulium
$4f^{14}6s^2$	6.254	173.04	12.18	$_{70}Yb\ ^1S_0$	25.05	43.6	Ytterbium
$4f^{14}5d6s^2$	5.426	174.97	13.9	$_{71}Lu\ ^2D_{3/2}$	20.96	45.25	Lutetium

Actinides and transuranides

Config		Mass		Z · Symbol · Term			Name
$6d^27s^2$	6.1	232.04	11.9	$_{90}Th\ ^3F_2$	18.3	28.7	Thorium
$5f^26d7s^2$	6.0	231.04	11.9	$_{91}Pa\ ^4K_{11/2}$			Protactinium
$5f^36d7s^2$	6.194	238.03	11.9	$_{92}U\ ^5L_6$	20	37	Uranium
$5f^46d7s^2$	6.266	237.05	11.9	$_{93}Np\ ^6L_{11/2}$			Neptunium
$5f^67s^2$	6.06	[244]		$_{94}Pu\ ^7F_0$			Plutonium
$5f^77s^2$	6.0	[243]		$_{95}Am\ ^8S_{7/2}$			Americium
$5f^{10}7s^2$	6.28	251		$_{98}Cf\ ^5I_8$			Californium
$5f^{12}7s^2$	6.50	257		$_{100}Fm\ ^3H_6$			Fermium
$5f^{14}7s^2$	6.63	259		$_{102}No\ ^1S_0$			Nobelium
$5f^{14}7s^27p$	4.90	266		$_{103}Lr\ ^2P_{1/2}$			Lawrencium
$5f^{14}6d^27s^2$	6.01	267		$_{104}Rf\ ^3F_2$			Rutherfordium
$5f^{14}6d^77s^2$	9.55	278		$_{109}Mt\ ^4F_{9/2}$			Meitnerium
$5f^{14}6d^87s^2$	10.38	281		$_{110}Ds\ ^3F_4$			Darmstadium
$5f^{14}6d^97s^2$	11.21	282		$_{111}Rg\ ^2D_{5/2}$			Roentgenium
$5f^{14}6d^{10}7s^2$	12.03	285		$_{112}Cn\ ^1S_0$			Copernicium
$5f^{14}6d^{10}7s^27p$	4.10	286		$_{113}Nh\ ^2P_{1/2}$			Nihonium
$5f^{14}6d^{10}7s^27p^2$	8.54	289		$_{114}Fl\ ^3P_0$			Flerovium
$5f^{14}6d^{10}7s^27p^3$	5.58	289		$_{115}Mc\ ^4S_{3/2}$			Moscovium
$5f^{14}6d^{10}7s^27p^4$	6.69	293		$_{116}Lv\ ^3P_2$			Livermorium
$5f^{14}6d^{10}7s^27p^5$	7.64	294		$_{117}Ts\ ^2P_{3/2}$			Tennessine
$5f^{14}6d^{10}7s^27p^6$	8.32	294		$_{118}Og\ ^1S_0$			Oganesson

Fig. 3.20 Ground states and ionization potentials of atoms and ions of lanthanides and transuranides

Electron

Group / Period	I	II	III	IV	V
1	$1s^2\,{}^1S_0$ ${}_1H$ 0.75420 Hydrogen				
2	$2s^2\,{}^1S_0$ ${}_3Li$ 0.61805 Lithium	$2p\,{}^2P_{1/2}$ ${}_4Be$ absent Beryllium	$2p^2\,{}^3P_0$ 0.27972 ${}_5B$ Boron	${}^4S_{3/2}$ 1.2621 $2p^3$ ${}_6C$ ${}^2D_{3/2}$ 0.033 Carbon	$2p^4\,{}^3P_0$ absent ${}_7N$ Nitrogen
3	$3s^2\,{}^1S_0$ ${}_{11}Na$ 0.54793 Sodium	$3p\,{}^2P_{1/2}$ ${}_{12}Mg$ absent Magnesium	$3p^2$ ${}^3P_0$ 0.4328 ${}_{13}Al$ ${}^1D_2$ 0.11 Aluminium	${}^4S_{3/2}$ 1.3895 $3p^3$ ${}^2D_{3/2}$ 0.5272 ${}_{14}Si$ ${}^2D_{5/2}$ 0.5255 Silicon ${}^2P_{1/2}$ 0.029	$3p^4\,{}^3P_0$ 0.7465 ${}_{15}P$ Phosphorus
4	$4s^2\,{}^1S_0$ ${}_{19}K$ 0.50146 Potassium	$4p\,{}^2P_{1/2}$ ${}_{20}Ca$ 0.0245 Calcium	$3d4s^24p$ ${}_{21}Sc$ ${}^3F_2$ 0.19 ${}^3D_1$ 0.04 Scandium	$3d^34s^2\,{}^4F_{3/2}$ ${}_{22}Ti$ 0.08 Titanium	$4d4s^2\,{}^5D_0$ ${}_{23}V$ 0.52 Vanadium
4	$3d^{10}4s^2\,{}^1S_0$ 1.23578 ${}_{29}Cu$ Copper	$4p\,{}^2P_{1/2}$ absent ${}_{30}Zn$ Zinc	$4p^2\,{}^3P_0$ 0.43 ${}_{31}Ga$ Gallium	${}^4S_{3/2}$ 1.2327 $4p^3$ ${}^2D_{3/2}$ 0.4014 ${}_{32}Ge$ ${}^2D_{5/2}$ 0.3773 Germanium	$4p^4\,{}^3P_0$ 0.81 ${}_{33}As$ Arsenic
5	$5s^2\,{}^1S_0$ ${}_{37}Rb$ 0.48592 Rubidium	$5p\,{}^2P_{1/2}$ ${}_{38}Sr$ 0.0520 Strontium	$4d5s^25p$ ${}_{39}Y$ ${}^3F_2$ 0.31 ${}^3D_1$ 0.16 Yttrium	$4d^35s^2\,{}^4F_{3/2}$ ${}_{40}Zr$ 0.43 Zirconium 52	${}_{41}Nb$ 0.89 Niobium
5	$4d^{10}5s^2\,{}^1S_0$ 1.30447 ${}_{47}Ag$ Silver	$5p\,{}^2P_{1/2}$ absent ${}_{48}Cd$ Cadmium	$5p^2\,{}^3P_0$ 0.40 In 49 Indium	${}^4S_{3/2}$ 1.1121 $5p^3$ ${}^2D_{3/2}$ 0.3976 ${}_{50}Sn$ ${}^2D_{5/2}$ 0.3046 Tin	${}^3P_2$ 1.0474 $5p^4$ ${}^3P_1$ 0.714 ${}^3P_0$ 0.700 ${}_{51}Sb$ ${}^1D_2$ 0.1308 Antimony
6	$6s^2\,{}^1S_0$ ${}_{55}Cs$ 0.47163 Cesium	$6p\,{}^2P_{1/2}$ ${}_{56}Ba$ 0.1446 Barium	$5d6s^2\,{}^3F_2$ ${}_{57}La$ 0.47 Lanthanum	$5d^36s^2\,{}^4F_{3/2}$ ${}_{58}Hf$ absent Hafnium	$5d^46s^2\,{}^5D_0$ ${}_{73}Ta$ 0.32 Tantalum
6	$5d^{10}6s^2\,{}^1S_0$ ${}_{79}Au$ 2.30863 Gold	$6p\,{}^2P_{1/2}$ absent ${}_{80}Hg$ Mercury	$6p^2\,{}^3P_0$ 0.38 ${}_{81}Tl$ Thallium	$6p^3\,{}^4S_{3/2}$ 0.364 ${}_{82}Pb$ Lead	$6p^4\,{}^3P_0$ 0.94236 ${}_{83}Bi$ Bismuth
7	$7s^2\,{}^1S_0$ ${}_{87}Fr$ 0.46 Francium				

Shell of valent electrons — Electron term

Symbol

Atomic number

Element

$3d^74s^2\,{}^4F_{9/2}$ ${}_{26}Fe$ 0.151 Iron

Electron binding energy

Fig. 3.21 Electron affinity of atoms

affinities

VI	VII	VIII		
		2s $^2S_{1/2}$ absent $_2$He Helium		
2p^5 $^2P_{3/2}$ 1.46111 $_8$O Oxygen	2p^6 $^1S_0$ 3.40119 $_9$F Fluorine	2p^{6}3s $^2S_{1/2}$ absent $_{10}$Ne Neon		
3p^5 $^2P_{3/2}$ 2.07710 $_{16}$S Sulfur	3p^6 $^1S_0$ 3.61269 $_{17}$Cl Chlorine	3p^{6}4s $^2S_{1/2}$ absent $_{18}$Ar Argon		
3d^{5}4s^2 $^6S_{5/2}$ $_{24}$Cr 0.6758 Chromium	3d^{6}4s^2 $^5D_4$ $_{25}$Mn absent Manganese	3d^{7}4s^2 $^4F_{9/2}$ $_{26}$Fe 0.151 Iron	3d^{8}4s^2 $^3F_4$ $_{27}$Co 0.663 Cobalt	3d^{9}4s^2 $^2D_{5/2}$ $_{28}$Ni 1.1572 Nickel
4p^5 $^2P_{3/2}$ 2.02067 $_{34}$Se Selenium	4p^6 $^1S_0$ 3.36359 $_{37}$Br Bromine	4p^{6}5s $^2S_{1/2}$ absent $_{36}$Kr Krypton		
4d^{5}5s^2 $^6S_{5/2}$ $_{42}$Mo 0.7472 Molibdenum	4d^{6}5s^2 $^5D_4$ $_{43}$Tc 0.6 Technetium	4d^{7}5s^2 $^4F_{9/2}$ $_{44}$Ru 1.0464 Ruthenium	4d^{8}5s^2 $^3F_4$ $_{45}$Rh 1.1429 Rhodium	4d^{10}5s $^2S_{1/2}$ 0.5621 $_{46}$Pd 4d^{9}5s^2 $^2D_{5/2}$ Palladium 0.4224
5p^5 $^2P_{3/2}$ 1.97087 $_{52}$Te Tellurium	5p^6 $^1S_0$ 3.05904 $_{53}$I Iodine	5p^{6}6s $^2S_{1/2}$ absent $_{54}$Xe Xenon		
5d^{5}6s^2 $^6S_{5/2}$ $_{74}$W 0.815 Tungsten	5d^{6}6s^2 $^5D_4$ $_{75}$Re 0.2 Rhenium	5d^{7}6s^2 $^4F_{9/2}$ $_{76}$Os 1.0778 Osmium	5d^{8}6s^2 $^3F_4$ $_{77}$Ir 1.5643 Iridium	5d^{9}6s^2 $^2D_{5/2}$ $_{78}$Pt 2.1251 Platinum
6p^5 $^2P_{3/2}$ 1.9 $_{84}$Po Polonium	6p^6 $^1S_0$ 2.8 $_{85}$At Astatine	6p^{6}7s $^2S_{1/2}$ absent $_{86}$Rn Radon		

Fig. 3.21 (continued)

Group \ Period	I	II	Lowest excited III	IV	V
1	$1s\,^2S_{1/2}$ n=2 10.199 n=3 12.088 n=4 12.479 $_1$H Hydrogen	$2s\,^3S_1$ 19.820 $1s^2$ $2s\,^1S_0$ 20.616 $^1S_0$ $2p\,^3P_2$ 20.964 $_2$He $2p\,^1P_1$ 21.218 Helium			
2	$2s\,^2S_{1/2}$ $2p\,^2P_{1/2}$ 1.848 $_3$Li $3s\,^2S_{1/2}$ 3.373 Lithium	$2s^2$ $2\,^3P_0$ 2.725 $_4$Be $2\,^1P_1$ 5.278 $^1S_0$ $2\,^3S_1$ 6.457 Berillium	$^4P_{1/2}$ 3.579 $2p$ $^2S_{1/2}$ 4.964 $_5$B $^2D_{3/2}$ 5.934 $2p\,^2P_{1/2}$ Boron	$2\,^1D_2$ 1.264 $2p^2$ $^3P_0$ $_6$C $2\,^1S_0$ 2.684 Carbon	$2\,^2D_{3/2}$ 2.384 $2p^3$ $^4S_{3/2}$ $_7$N $2\,^2P_{1/2}$3.576 Nitrogen
3	$3s\,^2S_{1/2}$ $3p\,^2P_{1/2}$ 2.102 $_{11}$Na $3p\,^2P_{3/2}$ 2.104 $4s\,^2S_{1/2}$ 2.607 Sodium	$3s^2\,^1S_0$ $3\,^3P_0$ 2.709 $_{12}$Mg $3\,^1P_1$ 4.316 $3\,^3S_1$ 5.108 Magnesium	$^2S_{1/2}$ 3.143 $3p$ $^4P_{1/2}$ 3.608 $^2P_{1/2}$ $^2D_{3/2}$ 4.022 $_{13}$Al $^2P_{1/2}$ 4.085 Aluminium	$^1D_2$ 0.781 $3p^2$ $^3P_0$ $_{14}$Si $^1S_0$ 1.909 Silicon	$^2D_{3/2}$ 1.409 $3p^3$ $^4S_{3/2}$ $_{15}$P $^2P_{1/2}$ 2.321 Phosphorus
4	$4s\,^2S_{1/2}$ $4p\,^2P_{1/2}$ 1.610 $_{19}$K $4p\,^2P_{3/2}$ 1.617 $5s\,^2S_{1/2}$ 2.607 Potassium f(4s-4p) = 1.05	$4s^2$ $^3P_0$ 1.879 $_{20}$Ca $^1P_1$ 2.521 $^1S_0$ $^1D_2$ 2.709 Calcium	$3d4s^2$ $^4F_{3/2}$ 1.439 $_{21}$Sc $^2F_{5/2}$ 1.859 $^2D_{3/2}$ $^4F_{3/2}$ 1.968 $^4D_{1/2}$ 1.985 Scandium	$3d^2 4s^2$ $^5F_1$ 0.813 $_{22}$Ti $^1D_2$ 0.900 $^3F_0$ 1.046 $^3F_2$ 1.430 Titanium	$3d^3 4s^2$ $^6D_{1/2}$ 0.262 $_{23}$V $^4D_{1/2}$ 1.043 $^4F_{3/2}$ $^4P_{1/2}$ 1.195 Vanadium
	$^2D_{5/2}$ 1.389 $3d^{10}4s\,^2S_{1/2}$ $^2D_{3/2}$ 1.642 $_{29}$Cu $^2P_{1/2}$ 3.786 Copper $^2P_{3/2}$ 3.817	$^3P_0$ 4.006 $4s$ $^1P_1$ 5.796 $_{30}$Zn $^3S_1$ 6.654 $^1S_0$ Zinc	$^2S_{1/2}$ 3.073 $4p$ $^2P_{1/2}$ $^2P_{1/2}$ $^2D_{3/2}$ 4.312 $_{31}$Ga $^2P_{1/2}$ 4.097 Gallium	$^1D_2$ 0.883 $4p^2$ $^1S_0$ 2.029 $^3P_0$ $^3P_0$ 4.643 $_{32}$Ge $^1P_1$ 4.962 Germanium	$^2D_{3/2}$ 1.313 $4p^3$ $^2P_{1/2}$ 2.255 $^4S_{3/2}$ 6.285 $_{33}$As Arsenic
5	$5s$ $5p\,^2P_{1/2}$ 1.560 $_{37}$Rb $5p\,^2P_{3/2}$ 1.589 $^2S_{1/2}$ $4d\,^2D_{3/2}$ 2.400 Rubidium	$5s^2$ $^3P_0$ 1.775 $_{38}$Sr $^3D_1$ 2.251 $^1S_0$ $^1D_2$ 2.498 Strontium $^1P_1$ 2.690	$4d5s^2$ $^2D_{5/2}$ 0.066 $_{39}$Y $^2P_{1/2}$ 1.306 $^2D_{3/2}$ $^4F_{3/2}$ 1.356 Yttrium	$4d^2 5s^2$ $^3F_3$ 0.071 $_{40}$Zr $^3F_4$ 0.154 $^3F_2$ $^3P_2$ 0.519 Zirconium	$4d^4 5s$ $^6D_{1/2}$ 0.142 $_{41}$Nb $^4P_{1/2}$ 0.620 Niobium
	$^2P_{1/2}$ 3.664 $4d^{10}5s$ $^2D_{3/2}$ 3.750 $_{47}$Ag $^2P_{3/2}$ 3.778 $^2S_{1/2}$ $^2D_{3/2}$ 4.304 Silver	$^3P_0$ 3.734 $5s^2$ $^1P_1$ 5.417 $_{48}$Cd $^3S_1$ 6.381 $^1S_0$ $^1S_0$ 6.610 Cadmium	$^2S_{1/2}$ 3.022 $5p$ $^2P_{1/2}$ $^2P_{1/2}$ $^2P_{1/2}$ 3.945 $_{49}$In $^2D_{3/2}$ 4.078 Indium	$^1D_2$ 1.068 $5p^2$ $^3P_0$ 2.128 $^3P_0$ $^3P_0$ 4.295 $_{50}$Sn $^1P_1$ 4.867 Tin	$^2D_{3/2}$ 1.055 $5p^3$ $^2P_{1/2}$ 2.033 $^4S_{3/2}$ $_{51}$Sb Antimony
6	$6s$ $6p\,^2P_{1/2}$ 1.386 $_{55}$Cs $6p\,^2P_{3/2}$ 1.455 $^2S_{1/2}$ $5d\,^2D_{3/2}$ 1.798 Cesium	$6s^2$ $^3D_1$ 1.120 $_{56}$Ba $^3P_0$ 1.521 $^1S_0$ $^1P_1$ 2.239 Barium	$5d6s^2$ $^2D_{5/2}$ 0.131 $_{57}$La $^2D_{3/2}$ 0.331 $^2F_{5/2}$ 0.869 $^4P_{1/2}$ 0.897 Lanthanum	$5d^2 6s^2$ $^3P_0$ 0.685 $_{72}$Hf $^1D_2$ 0.699 $^3F_2$ $^1G_4$ 1.306 Hafnium	$5d^3 6s^2$ $^4F_{5/2}$ 0.249 $_{73}$Ta $^4F_{3/2}$ $^4P_{1/2}$ 0.620 Tantalum
	$^2D_{5/2}$ 1.136 $5d^{10}6s$ $^2D_{3/2}$ 2.658 $_{79}$Au $^2P_{1/2}$ 4.632 $^2S_{1/2}$ $^2P_{3/2}$ 5.105 Gold	$^3P_0$ 4.667 $5d^{10}6s^2$ $^3P_1$ 4.887 $_{80}$Hg $^1S_0$ $^3P_2$ 5.461 $^1P_1$ 6.704 Mercury	$^2P_{3/2}$ 0.966 $6p$ $^2S_{1/2}$ 3.283 $^2P_{1/2}$ $^2P_{1/2}$ 4.235 $_{81}$Tl $^2P_{3/2}$ 4.359 Thallium	$^3P_1$ 0.970 $6p^2$ $^3P_2$ 1.320 $^3P_0$ $^1D_2$ 2.660 $_{82}$Pb $^1S_0$ 3.653 Lead	$^2D_{3/2}$ 1.416 $6p^3$ $^2P_{1/2}$ 2.686 $^4S_{3/2}$ $^4P_{1/2}$ 4.040 $_{83}$Bi Bismuth

Fig. 3.22 Lowest excited states of atoms

atom states

Legend

Shell of valence electrons: $4d^2 5s^2$ — Excitation energies in eV
Symbol: Zr
Atomic number: 40
Electron term of the ground state: $^3F_2$
Element: Zirconium
Electron terms of excited states: $^3F_3$ 0.071, $^3F_4$ 0.154, $^3P_2$ 0.519

Notations of LS-coupling scheme are used for electron terms.

VI	VII	VIII		
$^1D_2$ 1.967; $^1S_0$ 4.190 — $2p^4$, $^3P_2$, $_8$O, Oxygen	$^2P_{1/2}$ 0.050; 3 $^4P_{5/2}$ 12.717; 3 $^2P_{3/2}$ 12.999 — $2p^5$, $^2P_{3/2}$, $_9$F, Fluorine	$^3P_2$ 16.619; $^3P_1$ 16.671; $^3P_0$ 16.716; $^1P_1$ 16.848 — $2p^6$, $^1S_0$, $_{10}$Ne, Neon		
$^1D_2$ 1.145; $^1S_0$ 2.750 — $3p^4$, $^3P_2$, $_{16}$S, Sulfur	$^2P_{1/2}$ 0.109; $^4P_{5/2}$ 8.922; $^2P_{3/2}$ 9.282 — $3p^5$, $^2P_{3/2}$, $_{17}$Cl, Chlorine	$^3P_2$ 11.548; $^3P_1$ 11.624; $^3P_0$ 11.723; $^1P_1$ 11.828 — $3p^6$, $^1S_0$, $_{18}$Ar, Argon		
$3d^54s$, $^7S_3$; $^5S_2$ 0.941, $^5D_0$ 0.961 — $_{24}$Cr, Chromium	$3d^54s^2$, $^6S_{5/2}$; $^6D_{9/2}$ 2.114, $^8P_{5/2}$ 2.282 — $_{25}$Mn, Manganese	$3d^64s^2$, $^5D_4$; $^5F_5$ 0.859, $^3F_4$ 1.485, $^5P_3$ 2.176 — $_{26}$Fe, Iron	$3d^74s^2$, $^4F_{9/2}$; $^4F_{9/2}$ 0.432, $^2F_{7/2}$ 0.923 — $_{27}$Co, Cobalt	$3d^84s^2$, $^3F_4$; $^3D_3$ 0.025, $^1D_2$ 1.676, $^1S_0$ 1.826, $^3P_2$ 1.935 — $_{28}$Ni, Nickel
$^1D_2$ 1.187; $^1S_0$ 2.783; $^5S_2$ 5.974; $^5P_1$ 7.345 — $4p^4$, $^3P_2$, $_{34}$Se, Selenium	$^2P_{1/2}$ 0.457; $^4P_{5/2}$ 7.864; $^2P_{3/2}$ 8.329 — $4p^5$, $^2P_{3/2}$, $_{35}$Br, Bromine	$^3P_2$ 9.915; $^3P_1$ 10.032; $^3P_0$ 10.562; $^1P_1$ 10.644 — $4p^6$, $^1S_0$, $_{36}$Kr, Krypton		
$4d^55s$, $^7S_3$; $^5S_2$ 1.135, $^5D_0$ 1.360, $^5G_2$ 2.063, $^5P_3$ 2.260 — $_{42}$Mo, Molibdenum	$4d^55s^2$, $^6S_{5/2}$; $^6D_{9/2}$ 0.319, $^4D_{7/2}$ 1.304 — $_{43}$Tc, Technetium	$4d^75s$, $^5F_5$; $^5F_4$ 0.148, $^5F_3$ 0.259, $^5F_2$ 0.336, $^5F_1$ 0.385 — $_{44}$Ru, Ruthenium	$4d^85s$, $^4F_{9/2}$; $^2D_{5/2}$ 0.410, $^2F_{7/2}$ 0.693 — $_{45}$Rh, Rhodium	$4d^{10}$, $^1S_0$; $^3D_3$ 0.814, $^3D_2$ 0.962, $^3D_1$ 1.251, $^1D_2$ 1.453 — $_{46}$Pd, Palladium
$^3P_0$ 0.584; $^3P_1$ 0.589; $^1D_2$ 1.300; $^1S_0$ 2.876 — $5p^4$, $^3P_2$, $_{52}$Te, Tellurium	$^2P_{1/2}$ 0.943; $^4P_{5/2}$ 6.774; $^2P_{3/2}$ 7.665 — $5p^5$, $^2P_{3/2}$, $_{53}$I, Iodine	$^3P_2$ 8.315; $^3P_1$ 8.437; $^3P_0$ 9.447; $^1P_1$ 9.570 — $5p^6$, $^1S_0$, $_{54}$Xe, Xenon		
$5d^46s^2$, $^5D_0$; $^5D_1$ 0.207, $^7S_3$ 0.366, $^3P_0$ 1.181 — $_{74}$W, Tungsten	$5d^56s^2$, $^6S_{5/2}$; ^4P 1.436, $^6D_{9/2}$ 1.457, $^4G_{5/2}$ 1.813 — $_{75}$Re, Rhenium	$5d^66s^2$, $^5D_4$; $^5D_3$ 0.340, $^5D_2$ 0.516, $^5F_5$ 0.638 — $_{76}$Os, Osmium	$5d^76s^2$, $^4F_{9/2}$; $5d^86s\,^4F_{9/2}$ 0.352, $5d^76s^2\,^4F_{3/2}$ 0.508 — $_{77}$Ir, Iridium	$5d^96s$, $^3D_3$; $^3F_4$ 0.102, $^1S_0$ 0.761, $^3P_2$ 0.814 — $_{78}$Pt, Platinum

Fig. 3.22 (continued)

Splitting of lowest

Group / Period	I	II	III	IV	V
1	$1s\,^2S_{1/2}$ $2p(^2P_{1/2}-^2P_{3/2})$ 0.365 $2p\,^2P_{1/2}-2s\,^2S_{1/2}$ 0.035 $_1$H Hydrogen	$^3P_0-^3P_1$ 0.08 $^3P_0-^3P_2$ 1.06 $^1S_0$ $1s^2$ $_2$He Helium			
2	$2s\,^2S_{1/2}$ $^2P_{1/2}-^2P_{3/2}$ 0.3 $_3$Li Lithium	$2s^2$ $^1S_0$ $^3P_0-^3P_1$ 0.6 $^3P_0-^3P_2$ 3.0 $_4$Be Beryllium	$^2P_{1/2}-^2P_{3/2}$ 15.25 $2p\,^2P_{1/2}$ $_5$B Boron	$^3P_0-^3P_1$ 16.4 $^3P_0-^3P_2$ 43.4 $2p^2\,^3P_0$ $_6$C Carbon	$^2D_{3/2}-^2D_{5/2}$ 8.72 $2p^3\,^4S_{3/2}$ $_7$N Nitrogen
3	$3s\,^2S_{1/2}$ $^2P_{1/2}-^2P_{3/2}$ 17.2 $_{11}$Na Sodium	$3s^2\,^1S_0$ $^3P_0-^3P_1$ 20.05 $^3P_0-^3P_2$ 61.75 $_{12}$Mg Magnesium	$^2P_{1/2}-^2P_{3/2}$ 112.1 $^4P_{1/2}-^4P_{3/2}$ 46.6 $^4P_{1/2}-^4P_{5/2}$ 122.4 $3p\,^2P_{1/2}$ $_{13}$Al Aluminium	$^3P_0-^3P_1$ 77.1 $^3P_0-^3P_2$ 223.2 $3p^2\,^3P_0$ $_{14}$Si Silicon	$^2D_{3/2}-^2D_{5/2}$ 15.6 $3p^3\,^4S_{3/2}$ $_{15}$P Phosphorus
4	$4s\,^2S_{1/2}$ $^2P_{1/2}-^2P_{3/2}$ 57.7 $^2D_{5/2}-^2D_{3/2}$ 2.31 $_{19}$K Potassium	$4s^2\,^1S_0$ $^3P_0-^3P_1$ 52.16 $^3P_0-^3P_2$ 158.04 $_{20}$Ca Calcium	$3d4s^2$ $^2D_{3/2}-^2D_{5/2}$ 168.3 $^4F_{3/2}-^4F_{5/2}$ 37.7 $^2D_{3/2}$ $^4F_{3/2}-^4F_{7/2}$ 90.3 $^4F_{3/2}-^4F_{9/2}$ 157.4 $_{21}$Sc Scandium	$3d^24s^2$ $^3F_2-^3F_3$ 170.1 $^3F_2-^3F_4$ 386.9 $^3F_2$ $_{22}$Ti Titanium	$3d^34s^2$ $^4F_{3/2}-^4F_{5/2}$ 137.4 $^4F_{3/2}-^4F_{7/2}$ 323.5 $^4F_{3/2}-^4F_{9/2}$ 553.0 $^4F_{3/2}$ $_{23}$V Vanadium
4	$^2D_{5/2}-^2D_{3/2}$ 2042.9 $^2P_{1/2}-^2P_{3/2}$ 148.4 $3d^{10}4s\,^2S_{1/2}$ $_{29}$Cu Copper	$^3P_0-^3P_1$ 190.1 $^3P_0-^3P_2$ 579.0 $4s^2\,^1S_0$ $_{30}$Zn Zinc	$4p(^2P_{1/2}-^2P_{3/2})$ 826.2 $4p\,^2P_{1/2}$ $5p(^2P_{1/2}-^2P_{3/2})$ 110.9 $_{31}$Ga Gallium	$^3P_0-^3P_1$ 557.1 $^3P_0-^3P_2$ 1410.0 $4p^2\,^3P_0$ $_{32}$Ge Germanium	$^2D_{3/2}-^2D_{5/2}$ 323 $4p^3\,^4S_{3/2}$ $_{33}$As Arsenic
5	$5s\,^2S_{1/2}$ $^2P_{1/2}-^2P_{3/2}$ 237.6 $^2D_{5/2}-^2D_{3/2}$ 0.45 $_{37}$Rb Rubidium	$5s^2\,^1S_0$ $^3P_0-^3P_1$ 186.8 $^3P_0-^3P_2$ 581.0 $_{38}$Sr Strontium	$4d5s^2$ $^2D_{5/2}-^2D_{3/2}$ 530.4 $^2D_{3/2}$ $_{39}$Y Yttrium	$4d^25s^2$ $^3F_2-^3F_3$ 570.4 $^3F_2-^3F_4$ 1240.8 $^3F_2$ $_{40}$Zr Zirconium	$4d^45s$ $^6D_{1/2}-^6D_{3/2}$ 154.2 $^6D_{1/2}-^6D_{5/2}$ 392.0 $^6D_{1/2}-^6D_{7/2}$ 695.2 $^6D_{1/2}-^6D_{9/2}$ 1050.3 $^6D_{1/2}$ $_{41}$Nb Niobium
5	$^2D_{5/2}-^2D_{3/2}$ 4418 $^2P_{1/2}-^2P_{3/2}$ 920.7 $4d^{10}5s\,^2S_{1/2}$ $_{47}$Ag Silver	$^3P_0-^3P_1$ 542.1 $^3P_0-^3P_2$ 1713.0 $5s^2\,^1S_0$ $_{48}$Cd Cadmium	$^2P_{1/2}-^2P_{3/2}$ 2212.6 $5p\,^2P_{1/2}$ $_{49}$In Indium	$^3P_0-^3P_1$ 1692 $^3P_0-^3P_2$ 3428 $5p^2\,^3P_0$ $_{50}$Sn Tin	$^2D_{3/2}-^2D_{5/2}$ 1342 $5p^3\,^4S_{3/2}$ $_{51}$Sb Antimony
6	$6s\,^2S_{1/2}$ $^2P_{1/2}-^2P_{3/2}$ 554 $^2D_{3/2}-^2D_{5/2}$ 96.6 $_{55}$Cs Cesium	$6s^2$ $^3D_1-^3D_2$ 181.5 $^3D_1-^3D_3$ 562.6 $^1S_0$ $^3P_0-^3P_1$ 370.0 $^3P_0-^3P_2$ 1247.1 $_{56}$Ba Barium	$5d6s^2$ $^2D_{5/2}-^2D_{3/2}$ 1053.0 $^2D_{3/2}$ $_{57}$La Lanthanum	$5d^26s^2$ $^3F_2-^3F_3$ 2356.7 $^3F_2-^3F_4$ 4567.6 $^3F_2$ $_{72}$Hf Hafnium	$5d^36s^2$ $^4F_{3/2}-^4F_{5/2}$ 2010.1 $^4F_{3/2}-^4F_{7/2}$ 3963.9 $^4F_{3/2}-^4F_{9/2}$ 5621.0 $^4F_{3/2}$ $_{73}$Ta Tantalum
6	$^2D_{5/2}-^2D_{3/2}$ 12274 $^2P_{1/2}-^2P_{3/2}$ 3815 $5d^{10}6s\,^2S_{1/2}$ $_{79}$Au Gold	$^3P_0-^3P_1$ 1767.2 $^3P_0-^3P_2$ 6397.9 $5d^{10}6s^2\,^1S_0$ $_{80}$Hg Mercury	$^2P_{1/2}-^2P_{3/2}$ 7793 $6p\,^2P_{1/2}$ $_{81}$Tl Thallium	$^3P_0-^3P_1$ 7819.3 $^3P_0-^3P_2$ 10650.3 $6p^2\,^3P_0$ $_{82}$Pb Lead	$^2D_{3/2}-^2D_{5/2}$ 4019 $6p^3\,^4S_{3/2}$ $_{83}$Bi Bismuth

Fig. 3.23 Splitting of lowest atom levels

atom levels

Legend (top right):

difference of energies for comparable these states / Shell of valence electrons energy levels in cm⁻¹

- Symbol → $4d^8 5s$ | $^4F_{9/2}-^4F_{7/2}$ 1530.0
- Atomic number → $_{45}$ **Rh** | $^4F_{9/2}-^4F_{5/2}$ 2598.0
- Electron term of the ground state → $^4F_{9/2}$ | $^4F_{9/2}-^4F_{3/2}$ 3472.7
- Rhodium (Element)

Group VI

Oxygen — $2p^4$, $^3P_2-^3P_1$ 158.3, $^3P_2$ $_8$**O**, $^3P_2-^3P_0$ 227.0

Sulfur — $3p^4$, $^3P_2-^3P_1$ 396.0, $^3P_2$ $_{16}$**S**, $^3P_2-^3P_0$ 573.6

Chromium — $3d^5 4s$, $^5D_0-^5D_1$ 60.0, $^5D_0-^5D_2$ 176.7, $_{24}$**Cr**, $^5D_0-^5D_3$ 344.4, $^7S_3$, $^5D_0-^5D_4$ 556.8

Selenium — $4p^4$, $^3P_2-^3P_1$ 1989.5, $^3P_2$ $_{34}$**Se**, $^3P_2-^3P_0$ 2534.4

Molibdenum — $4d^5 5s$, $^5D_0-^5D_1$ 176.8, $_{42}$**Mo**, $^5D_0-^5D_2$ 892.5, $^7S_3$, $^5D_0-^5D_3$ 1480.3, $^5D_0-^5D_4$

Tellurium — $5p^4$, $^3P_2-^3P_1$ 4706.5, $^3P_2$ $_{52}$**Te**, $^3P_2-^3P_0$ 4750.7

Tungsten — $5d^4 6s^2$, $^5D_0-^5D_1$ 1670.1, **W**, $^5D_0-^5D_2$ 3325.5, $_{74}$ $^5D_0$, $^5D_0-^5D_3$ 4830.0, $^7S_3$, $^5D_0-^5D_4$ 6219.3

Group VII

Fluorine — $2p^5$, $^2P_{3/2}-^2P_{1/2}$ 404.1, $^2P_{3/2}$ $_9$**F**

Chlorine — $3p^5$, $^2P_{3/2}-^2P_{1/2}$ 882.4, $^2P_{3/2}$ $_{17}$**Cl**

Manganese — $3d^5 4s^2$, $^6D_{9/2}-^6D_{7/2}$ 229.7, $^6D_{9/2}-^6D_{5/2}$ 399.2, $_{25}$**Mn**, $^6D_{9/2}-^6D_{3/2}$ 516.2, $^6S_{5/2}$, $^6D_{9/2}-^6D_{1/2}$ 584.8

Bromine — $4p^5$, $^2P_{3/2}-^2P_{1/2}$ 3685, $^2P_{3/2}$ $_{35}$**Br**

Technetium — $4d^5 5s^2$, $^6D_{9/2}-^6D_{7/2}$ 678.0, $^6D_{9/2}-^6D_{5/2}$ 1129.6, $_{43}$**Tc**, $^6D_{9/2}-^6D_{3/2}$ 1429.7, $^6S_{5/2}$, $^6D_{9/2}-^6D_{1/2}$ 1605.8

Iodine — $5p^5$, $^2P_{3/2}-^2P_{1/2}$ 7603.1, $^2P_{3/2}$ $_{53}$**I**

Rhenium — $5d^5 6s^2$, $^4P_{5/2}-^4P_{3/2}$ 2242.1, $_{75}$**Re**, $^6S_{5/2}$, $^6D_{9/2}-^6D_{7/2}$ 2462.4

Group VIII

Neon — $2p^6$, $^3P_2-^3P_1$ 417.4, $^3P_2-^3P_0$ 776.8, $_{10}$**Ne**, $^3P_2-^1P_1$ 1846.9, $^1S_0$

Argon — $2p^6$, $^3P_2-^3P_1$ 606.8, $^1S_0$, $^3P_2-^3P_0$ 1409.9, $_{18}$**Ar**, $^3P_2-^1P_1$ 2256.1

Iron — $3d^6 4s^2$, $^5D_4-^5D_3$ 415.9, $^5D_4-^5D_2$ 704.0, $_{26}$**Fe**, $^5D_4-^5D_1$ 888.1, $^5D_4$, $^5D_4-^5D_0$ 978.1

Krypton — $4p^6$, $^3P_2-^3P_1$ 945.0, $^1S_0$, $^3P_2-^3P_0$ 5219.9, $_{36}$**Kr**, $^3P_2-^1P_1$ 5875.0

Ruthenium — $4d^7 5s$, $^5F_5-^5F_4$ 1190.6, $^5F_5-^5F_3$ 2091.5, $_{44}$**Ru**, $^5F_5-^5F_2$ 2713.2, $^5F_5$, $^5F_5-^5F_1$ 3105.5

Xenon — $5p^6$, $^3P_2-^3P_1$ 977.6, $^1S_0$, $^3P_2-^3P_0$ 9129.2, $_{54}$**Xe**, $^3P_2-^1P_1$ 10117.5

Osmium — $5d^6 6s^2$, $^5D_4-^5D_3$ 2740.5, $^5D_4-^5D_2$ 4159.3, $_{76}$**Os**, $^5D_4$, $^5D_4-^5D_1$ 5766.1, $^5D_4-^5D_0$ 6092.8

Group VIII (continued)

Cobalt — $3d^7 4s^2$, $^4F_{9/2}-^4F_{7/2}$ 816.0, $^4F_{9/2}-^4F_{5/2}$ 1406.8, $_{27}$**Co**, $^4F_{9/2}$, $^4F_{9/2}-^4F_{3/2}$ 1809.3

Rhodium — $4d^8 5s$, $^4F_{9/2}-^4F_{7/2}$ 1530.0, $^4F_{9/2}-^4F_{5/2}$ 2598.0, $_{45}$**Rh**, $^4F_{9/2}$, $^4F_{9/2}-^4F_{3/2}$ 3472.7

Iridium — $5d^7 6s^2$, $^4F_{9/2}-^4F_{9/2}$ 2835.0, $_{77}$**Ir**, $^4F_{9/2}$

Nickel — $3d^8 4s^2$, $^3F_4-^3F_3$ 1332.2, $^3F_4-^3F_2$ 2216.5, $_{28}$**Ni** $^3F_4$, $^3D_3-^3D_2$ 675.0, $^3D_3-^3D_1$ 1508.3

Palladium — $4d^{10}$, $^3D_3-^3D_2$ 1191, $_{46}$**Pd** $^1S_0$, $^3D_3-^3D_1$ 3530, $^3D_3-^1D_2$ 5158

Platinum — $5d^9 6s$, $^3D_3-^3D_2$, $_{78}$**Pt** $^3D_3$, 776

Fig. 3.23 (continued)

Tables

Table 3.1 The angle electron wave function in the hydrogen atom for small electron momenta l [44–46]

l	m	$Y_{lm}(\theta, \varphi)$
0	0	$\frac{1}{\sqrt{4\pi}}$
1	0	$\sqrt{\frac{3}{4\pi}} \cdot \cos\theta$
1	± 1	$\pm\sqrt{\frac{3}{8\pi}} \cdot \sin\theta \cdot \exp(\pm i\varphi)$
2	0	$\sqrt{\frac{5}{4\pi}} \cdot (\frac{3}{2}\cos^2\theta - \frac{1}{2})$
2	± 1	$\pm\sqrt{\frac{15}{8\pi}} \cdot \sin\theta \cdot \cos\theta \cdot \exp(\pm i\varphi)$
2	± 2	$\frac{1}{2}\sqrt{\frac{15}{8\pi}} \cdot \sin^2\theta \cdot \exp(\pm 2i\varphi)$
3	0	$\sqrt{\frac{7}{4\pi}} \cdot (5\cos^3\theta - 3\cos\theta)$
3	± 1	$\pm\frac{1}{8}\sqrt{\frac{21}{\pi}} \cdot \sin\theta \cdot (5\cos^2\theta - 1)\exp(\pm i\varphi)$
3	± 2	$\frac{1}{4}\sqrt{\frac{105}{2\pi}} \cdot \sin^2\theta\cos\theta \cdot \exp(\pm 2i\varphi)$
3	± 3	$\pm\frac{1}{8}\sqrt{\frac{35}{\pi}} \cdot \sin^3\theta \cdot \exp(\pm 3i\varphi)$
4	0	$\sqrt{\frac{9}{4\pi}} \cdot \left(\frac{35}{8}\cos^4\theta - \frac{15}{4}\cos^2\theta + \frac{3}{8}\right)$
4	± 1	$\pm\frac{3}{8}\sqrt{\frac{5}{\pi}} \cdot \sin\theta \cdot (7\cos^3\theta - 3\cos\theta) \cdot \exp(\pm i\varphi)$
4	± 2	$\pm\frac{3}{8}\sqrt{\frac{5}{2\pi}} \cdot \sin^2\theta \cdot (7\cos^2\theta - 1) \cdot \exp(\pm 2i\varphi)$
4	± 3	$\pm\frac{3}{8}\sqrt{\frac{35}{\pi}} \cdot \sin^3\theta \cdot \cos\theta \cdot \exp(\pm 3i\varphi)$
4	± 4	$\frac{3}{16}\sqrt{\frac{35}{2\pi}} \cdot \sin^4\theta \cdot \exp(\pm 4i\varphi)$

Table 3.2 Electron radial wave functions R_{nl} for the hydrogen atom [44–46]

State	R_{nl}
$1s$	$2\exp(-r)$
$2s$	$\frac{1}{\sqrt{2}}\left(1 - \frac{r}{2}\right)\exp(-r/2)$
$2p$	$\frac{1}{\sqrt{24}}r\exp(-r/2)$
$3s$	$\frac{2}{3\sqrt{3}}\left(1 - \frac{2r}{3} + \frac{2r^2}{27}\right)\exp(-r/3)$
$3p$	$\frac{2}{27}\sqrt{\frac{2}{3}}\cdot r\left(1 - \frac{r}{6}\right)\exp(-r/3)$
$3d$	$\frac{4}{81\sqrt{30}}\cdot r^2\exp(-r/3)$
$4s$	$\frac{1}{4}\left(1 - \frac{3r}{4} + \frac{r^2}{8} - \frac{r^3}{192}\right)\exp(-r/4)$
$4p$	$\frac{1}{16}\sqrt{\frac{5}{3}}\cdot r\cdot\left(1 - \frac{r}{4} + \frac{r^2}{80}\right)\exp(-r/4)$
$4d$	$\frac{1}{64\sqrt{5}}\cdot r^2\cdot\left(1 - \frac{r}{12}\right)\exp(-r/4)$
$4f$	$\frac{1}{768\sqrt{35}}\cdot r^3\cdot\exp(-r/4)$

Table 3.3 Formulas for average values $\langle r^n\rangle$ given in atomic units, where r is the electron distance from the Coulomb center [45]

Value	Formula
$\langle r\rangle$	$\frac{1}{2}\cdot\left[3n^2 - l(l+1)\right]$
$\langle r^2\rangle$	$\frac{n^2}{2}\cdot\left[5n^2 + 1 - 3l(l+1)\right]$
$\langle r^3\rangle$	$\frac{n^2}{8}\cdot\left[35n^2(n^2 - 1) - 30n^2(l+2)(l-1) + 3(l+2)(l+1)l(l-1)\right]$
$\langle r^4\rangle$	$\frac{n^4}{8}\cdot\left[63n^4 - 35n^2(2l^2 + 2l - 3) + 5l(l+1)(3l^2 + 3l - 10) + 12\right]$
$\langle r^{-1}\rangle$	n^{-2}
$\langle r^{-2}\rangle$	$[n^3(l + 1/2)]^{-1}$
$\langle r^{-3}\rangle$	$[n^3(l+1)\cdot(l + 1/2)\cdot l]^{-1}$
$\langle r^{-4}\rangle$	$[3n^2 - l(l+1)][2n^5\cdot(l + 3/2)\cdot(l+1)\cdot(l + 1/2)\cdot l\cdot(l - 1/2)]^{-1}$

Table 3.4 Average values $\langle r^n\rangle$ for lowest states of the hydrogen atom expressed in atomic units [45]

State	$\langle r\rangle$	$\langle r^2\rangle$	$\langle r^3\rangle$	$\langle r^4\rangle$	$\langle r^{-1}\rangle$	$\langle r^{-2}\rangle$	$\langle r^{-3}\rangle$	$\langle r^{-4}\rangle$
$1s$	1.5	3	7.5	22.5	1	2	–	–
$2s$	6	42	330	2880	0.25	0.25	–	–
$2p$	5	30	210	1680	0.25	0.0833	0.0417	0.0417
$3s$	13.5	207	3442	$6.136\cdot10^4$	0.111	0.0741	–	–
$3p$	12.5	180	2835	$4.420\cdot10^4$	0.111	0.0247	0.0123	0.0137
$3d$	10.5	126	1701	$2.552\cdot10^4$	0.111	0.0148	0.0247	$5.49\cdot10^{-3}$
$4s$	24	648	18720	$5.702\cdot10^5$	0.0625	0.0312	–	–
$4p$	23	600	16800	$4.973\cdot10^5$	0.0625	0.0104	$5.21\cdot10^{-3}$	$5.49\cdot10^{-4}$
$4d$	21	504	13100	$3.629\cdot10^5$	0.0625	0.00625	$1.04\cdot10^{-3}$	$2.60\cdot10^{-4}$
$4f$	18	360	7920	$1.901\cdot10^5$	0.0625	0.00446	$3.72\cdot10^{-4}$	$3.7\cdot10^{-5}$

Table 3.5 Quantum defect for atoms of alkali metals [50, 51]

Atom	Li	Na	K	Rb	Cs
δ_s	0.399	1.347	2.178	3.135	4.057
δ_p	0.053	0.854	1.712	2.65	3.58
δ_d	0.002	0.0145	0.267	1.34	2.47
δ_f	–	0.0016	0.010	0.0164	0.033

Table 3.6 Dependence of interactions of heliumlike ions on the nuclear charge Z

Parameter	Z-dependence
Ionization potential	Z^2
Potential of exchange interaction	Z
Spin-orbit interaction	Z^4
Rate of one-photon radiative transition	Z^4
Rate of two-photon radiative transition	Z^8

Table 3.7 Parameters of the lowest states for atoms of the third group of the periodic system of elements with one p-valence electron

Atom	B	Al	Ga	In	Tl
Ground shell	2p	3p	4p	5p	6p
Excited shell	3s	4s	5s	6s	7s
ε_{ex}, eV	4.96	3.14	3.07	3.02	3.28
$\Delta\varepsilon(^2P_{1/2} - ^2P_{3/2})$, cm^{-1}	15.2	112	896	2213	7793
Z_{ef}	2.5	4.2	7.0	8.8	12
$\lambda(^2P_{1/2} - ^2S_{1/2})$, nm	249.68	394.40	403.30	410.18	377.57
$\tau(^2S_{1/2})$, ns	3.6	6.8	6.2	7.4	7.6

Table 3.8 Energetic parameters of the lowest states for atoms of the fourth group of the periodic system of elements

Atom	C	Si	Ge	Sn	Pb
Shell	$2p^2$	$3p^2$	$4p^2$	$5p^2$	$6p^2$
$\varepsilon_{ex}(^1D)$, eV	1.26	0.78	0.88	1.07	2.66
$\varepsilon_{ex}(^1S)$, eV	2.68	1.91	2.03	2.13	3.65
$\varepsilon_{ex}(^3P_1)$, cm^{-1}	16.4	77.1	557	1692	7819
$\varepsilon_{ex}(^3P_2)$, cm^{-1}	43.4	223	1410	3428	10650

Table 3.9 Energetic parameters of the lowest states for atoms of the fifth group of the periodic system of elements

Atom	N	P	As	Sb	Bi
Shell	$2p^3$	$3p^3$	$4p^3$	$5p^3$	$6p^3$
$\varepsilon_{ex}(^2D_{5/2})$, eV	2.38	1.41	1.31	1.06	1.42
$\varepsilon_{ex}(^2P_{1/2})$, eV	3.58	2.32	2.25	2.03	2.68
$\Delta\varepsilon(^2D_{5/2} -^2 D_{3/2})$, cm^{-1}	8.7	15.6	322	1342	4019
$\Delta\varepsilon(^2P_{1/2} -^2 P_{3/2})$, cm^{-1}	0.39	25.3	461	2069	10927

Table 3.10 Energetic parameters of the lowest states for atoms of the sixth group of the periodic system of elements

Atom	O	S	Se	Te
Shell	$2p^4$	$3p^4$	$4p^4$	$5p^4$
$\varepsilon_{ex}(^1D)$, eV	1.97	1.14	1.19	1.31
$\varepsilon_{ex}(^1S)$, eV	4.19	2.75	2.78	2.88
$\varepsilon_{ex}(^3P_1)$, cm^{-1}	158	396	1990	4751
$\varepsilon_{ex}(^3P_2)$, cm^{-1}	227	574	2534	4706

Table 3.11 Energetic parameters for halogen atoms

Atom	F	Cl	Br	I
Shell	$2p^5$	$3p^5$	$4p^5$	$5p^5$
J, eV	17.42	12.97	11.81	10.45
Lowest excited term	$2p^43s,^4P_{5/2}$	$3p^44s,^4P_{5/2}$	$4p^45s,^4P_{5/2}$	$5p^46s,^4P_{5/2}$
$\varepsilon_{ex}(^4P_{5/2})$, eV	12.70	8.92	7.86	6.77
$\Delta\varepsilon(^2P_{3/2} -^2 P_{1/2})$, cm^{-1}	404	881	3685	7603

Table 3.12 Fractional parentage coefficients for electrons of s and p shells [8, 57]

Atom	Atomic rest	$G^{LS}_{L'S'}$	Atom	Atomic rest	$G^{LS}_{L'S'}$
$s(^2S)$	(^1S)	1	$p^3(^2P)$	$p^2(^1S)$	$\sqrt{2/3}$
$s^2(^1S)$	$s(^2S)$	1	$p^4(^3P)$	$p^3(^4S)$	$-1/\sqrt{3}$
$p(^2P)$	(^1S)	1		$p^3(^2D)$	$\sqrt{5/12}$
$p^2(^3P)$	$p(^2P)$	1		$p^3(^2P)$	$-1/2$
$p^2(^1D)$	$p(^2P)$	1	$p^4(^1D)$	$p^3(^4S)$	0
$p^2(^1S)$	$p(^2P)$	1		$p^3(^2D)$	$\sqrt{3/4}$
$p^3(^4S)$	$p^2(^3P)$	1		$p^3(^2P)$	$-1/2$
	$p^2(^1D)$	0	$p^4(^1S)$	$p^3(^4S)$	0
	$p^2(^1S)$	0		$p^3(^2D)$	0
$p^3(^2D)$	$p^2(^3P)$	$1/\sqrt{2}$		$p^3(^2P)$	1
	$p^2(^1D)$	$-1/\sqrt{2}$	$p^5(^2P)$	$p^4(^3P)$	$\sqrt{3/5}$
	$p^2(^1S)$	0		$p^4(^1D)$	$1/\sqrt{3}$
$p^3(^2P)$	$p^2(^3P)$	$-1/\sqrt{2}$		$p^4(^1S)$	$1/\sqrt{15}$
	$p^2(^1D)$	$-\sqrt{5/18}$	$p^6(^1S)$	$p^5(^2P)$	1

Table 3.13 States of atoms with non-filled electron shells

Shell configuration	Number of levels	Number of terms	Statistical weight
s	1	1	2
s^2	1	1	1
p, p^5	1	2	6
p^2, p^4	3	5	15
p^3	3	5	20
d, d^9	1	2	10
d^2, d^8	5	9	45
d^3, d^7	8	19	120
d^4, d^6	18	40	210
d^5	16	37	252
f, f^{13}	1	2	14
f^2, f^{12}	7	13	91
f^3, f^{11}	17	41	364
f^4, f^{10}	47	107	1001
f^5, f^9	73	197	2002
f^6, f^8	119	289	3003
f^7	119	327	3432

Table 3.14 Electron terms of atoms with filling electron shells d^n [19]

n	Electron terms	Number of states	Number of electron terms
0, 10	^1S	1	1
1, 9	^2D	10	2
2, 8	$^1S, ^3P, ^1D, ^3F, ^1G$	45	9
3, 7	$^2P, ^4P, ^2D(2), ^2F, ^4F, ^2G, ^2H$	120	19
4, 6	$^1S(2), ^3P(4), ^1D(2), ^3D, ^5D,$		
	$^1F, ^3F(2), ^1G(2), ^3G, ^3H, ^1J$	210	40
5	$^2S, ^6S, ^2P, ^4P, ^2D(3), ^2F(2),$		
	$^4F, ^2G(2), ^4G, ^2H, ^2J$	252	37

Table 3.15 The ground states for atoms with filling d-shell [19]

Electron shell	Term of the ground state	Atoms with this electron shell
d	$^2D_{3/2}$	$Sc(3d), Y(4d), La(5d), Lu(5d), Ac(6d), Lr(6d)$
d^2	$^3F_2$	$Ti(3d^2), Zr(4d^2), Hf(5d^2), Th(6d^2)$
d^3	$^4F_{3/2}$	$V(3d^3), Ta(5d^3)$
d^4	$^5D_0$	$W(5d^4)$
d^5	$^6S_{5/2}$	$Mn(3d^5), Tc(4d^5), Re(5d^5)$
d^6	$^5D_4$	$Fe(3d^6), Os(5d^6)$
d^7	$^4F_{9/2}$	$Co(3d^7), Ir(5d^7)$
d^8	$^3F_4$	$Ni(3d^8)$
d^9	$^2D_{5/2}$	–

Table 3.16 The ground state of atoms with a filling f-shell [19]

Electron shell	Term of the ground state	Atoms with this electron shell
f	$^2F_{5/2}$	–
f^2	$^3H_4$	–
f^3	$^4I_{9/2}$	$Pr(4f^3)$
f^4	$^5I_4$	$Nd(4f^4)$
f^5	$^6H_{5/2}$	$Pm(4f^5)$
f^6	$^7F_0$	$Sm(4f^6), Pu(5f^6)$
f^7	$^8S_{7/2}$	$Eu(4f^7), Am(5f^7)$
f^8	$^7F_6$	–
f^9	$^6H_{15/2}$	$Tb(4f^9), Bk(5f^9)$
f^{10}	$^5I_8$	$Dy(4f^{10}), Cf(5f^{10})$
f^{11}	$^4I_{15/2}$	$Ho(4f^{11}), Es(5f^{11})$
f^{12}	$^3H_6$	$Er(4f^{12}), Fm(5f^{12})$
f^{13}	$^2F_{5/2}$	$Tm(4f^{13}), Md(5f^{13})$

Table 3.17 Electron terms for filling of electron shells with valence p-electrons

LS-scheme	LS-term	J	jj-scheme	J
p	^2P	1/2	$[1/2]^1$	1/2
	^2P	3/2	$[3/2]^1$	3/2
p^2	^3P	0	$[1/2]^2$	0
	^3P	1	$[1/2]^1[3/2]^1$	1
	^3P	2	$[1/2]^1[3/2]^1$	2
	^1D	2	$[3/2]^2$	2
	^1S	0	$[3/2]^2$	0
p^3	^4S	3/2	$[1/2]^2[3/2]^1$	3/2
	^2D	3/2	$[1/2]^1[3/2]^2$	3/2
	^2D	5/2	$[1/2]^1[3/2]^2$	5/2
	^2P	1/2	$[1/2]^1[3/2]^2$	1/2
	^2P	3/2	$[3/2]^3$	3/2
p^4	^3P	2	$[1/2]^1[3/2]^3$	2
	^3P	0	$[1/2]^2[3/2]^2$	0
	^3P	1	$[1/2]^{21}[3/2]^2$	1
	^1D	2	$[1/2]^1[3/2]^3$	2
	^1S	0	$[3/2]^4$	0
p^5	^2P	3/2	$[1/2]^2[3/2]^3$	3/2
	^2P	1/2	$[1/2]^1[3/2]^4$	1/2
p^6	^1S	0	$[1/2]^2[3/2]^4$	0

Table 3.18 Lower excited states of atoms of inert gases and parameters of the scheme (3.25)

Atom	Δ_f, cm^{-1}	$\varepsilon_3 - \varepsilon_5$, cm^{-1}	$(\varepsilon_3 - \varepsilon_5)/\Delta_f$	b, cm^{-1}	a/b	$(\varepsilon_2 - \varepsilon_4)/C$	$(\varepsilon_2 - \varepsilon_4)/D$
Ne	780.3	776.8	0.996	1482	0.35	1.00	1.00
Ar	1432.0	1410	0.985	1453	0.66	0.99	1.00
Kr	5370.1	5214	0.972	1594	2.24	0.97	1.00
Xe	10537	9129	0.866	1966	3.57	1.13	1.05

Chapter 4
Rates of Radiative Transitions and Atomic Spectra

Abstract The strongest radiative transitions due to dipole interaction are considered. Selection rules are analyzed for atoms. As an example, radiative transitions are represented with participation of lowest excited states of inert gas atoms. Broadening of spectral lines may be resulted from various mechanisms including Doppler broadening, Lorenz broadening, and also quasistatic shift and broadening of spectral lines. Parameters of atom radiation and absorption are defined and analyzed.

4.1 Dipole Radiation of Atomic Particles

Radiative transitions between states of atomic particles result from interaction of an electromagnetic wave and atomic particle. The strongest radiative transitions proceed due to dipole interaction of these systems, and the operator of this interaction has the form

$$V = -\mathbf{E}\mathbf{D}, \tag{4.1}$$

where $\mathbf{E}$ is the electric field strength of radiation electromagnetic field, and $\mathbf{D}$ is the dipole moment operator for an atomic particles. Let us consider first the radiative transition process between discrete states o and f of an atomic particle A which proceeds according to the scheme

$$A_o + \hbar\omega \rightarrow A_* \tag{4.2}$$

The energy conservation law is fulfilled for radiative processes

$$\varepsilon_o + \hbar\omega = \varepsilon_*, \tag{4.3}$$

where ε_o and ε_* are the energies of the initial and final states of an atomic particle, $\hbar\omega$ is the energy of an absorbed photon.

The probability of the dipole radiative transition per unit time is equal [56, 57, 65]

$$w_{of} = \frac{4\omega^3}{3\hbar c^3} \cdot n_\omega \cdot |\langle o \, |\mathbf{D}| \, * \rangle|^2 \, g_* n_\omega, \tag{4.4}$$

© Springer International Publishing AG, part of Springer Nature 2018

B. M. Smirnov, *Atomic Particles and Atom Systems*, Springer Series on Atomic, Optical, and Plasma Physics 51, https://doi.org/10.1007/978-3-319-75405-5_4

where n_ω is a number of photons of a given frequency, g_* is the statistical weight of the final state of an atomic particle, and this expression is averaged over polarizations of an electromagnetic wave. From this it follows the expression for the radiative lifetime τ_{*o} for a state of an atomic particle with respect to radiative transition into a state o

$$\frac{1}{\tau_{*o}} = \frac{4\omega^3}{3\hbar c^3} |\langle o |\mathbf{D}| *\rangle|^2 g_o \tag{4.5}$$

Table 4.1 contains the conversional factors in formulas for photon parameters and radiative lifetime for radiative dipole transitions.
    Explanation to Table 4.1.
    1. The photon energy $\varepsilon = \hbar\omega$, where ω is the photon frequency.
    2. The photon frequency is $\omega = \varepsilon/\hbar$.
    3. The photon frequency is $\omega = 2\pi c/\lambda$, where λ is the wavelength, c is the light speed.
    4. The photon energy is $\varepsilon = 2\pi\hbar c/\lambda$.
    5. The oscillator strength for a radiative transition from the lower o to the upper $*$ state of an atomic particle that is averaged over lower states o and is summed over upper states $*$ is equal to

$$f_{o*} = \frac{2m_e\omega}{3\hbar e^2} |\langle o |\mathbf{D}| *\rangle|^2 g_* = \frac{2m_e\omega}{3\hbar e^2} d^2 g_*, \tag{4.6}$$

where $\mathbf{d} = \langle o|\mathbf{D}|*\rangle$ is the matrix element for the dipole moment operator of an atomic particle taken between transition states. Here m_e, $\hbar$ are atomic parameters, g_* is the statistical weight of the upper state, $\omega = (\varepsilon_* - \varepsilon_o)/\hbar$ is the transition frequency, where $\varepsilon_o, \varepsilon_*$ are the energies of transition states.
    6. The oscillator strength for radiative transition is [21, 45]

$$f_{o*} = \frac{4\pi cm_e}{3\hbar e^2\lambda} d^2 g_* \tag{4.7}$$

Here λ is the transition wavelength; other notations are the same as above.
    7. The rate of the radiative transition is [56, 57, 65]

$$\frac{1}{\tau_{*o}} = B_{*o} = \frac{4\omega^3}{3\hbar c^3} d^2 g_o \tag{4.8}$$

Here B is the Einstein coefficient; other notations are as above.
    8. The rate of radiative transition is given by

$$\frac{1}{\tau_{*o}} = B_{*o} = \frac{32\pi^3}{3\hbar\lambda^3} d^2 g_o \tag{4.9}$$

Here λ is the wavelength of this transition; other notations are the same as above.

9. The rate of radiative transition is

$$\frac{1}{\tau_{*o}} = \frac{2\omega^2 e^2 g_o}{m_e c^3 g_*} f_{o*} \qquad (4.10)$$

10. The rate of radiative transition is

$$\frac{1}{\tau_{*o}} = \frac{8\pi^2 g_o}{\hbar g_* \lambda^2 c} f_{o*} \qquad (4.11)$$

4.2 Selection Rules for Radiative Transitions in Atomic Particles

The strongest radiative transitions between discrete atom states in a non-relativistic case connects atom states with a nonzero matrix elements of the dipole moment operator. This follows from the above formulas for the oscillator strength of radiative transitions and leads to certain selection rules for the radiative transitions. For transitions between hydrogen atom states when one electron is located in the Coulomb field of a charged center these selection rules have the form [21, 45]

$$\Delta l = \pm 1, \quad \Delta m = 0, \pm 1 \qquad (4.12)$$

Table 4.2 contains parameters of radiative transitions for the hydrogen atom.

Lower excited states with the possibility of dipole radiative transitions in the ground atom state are called the resonantly excited states. Positions of levels and parameters of radiative transitions to the ground atom states are contained in the periodical table of Fig. 4.1. Radiative transitions between states with zero matrix element of the dipole moment operator between these states are weaker than those for dipole radiation transitions, and such transitions are named as forbidden radiative transitions. As the transition energy increases, the difference in radiative transitions for resonant and forbidden transitions decreases. This is demonstrated by data of Table 4.3 where the times of identical radiative transitions are compared for helium-like ions.

The development of physics of atoms and molecules in a most degree is connected with spectroscopy of atoms and molecules because of a high resolution of spectral lines together with other measurements which give detailed information about atoms and molecules (for example, [66–70]. This information includes also radiative properties of atoms in various states.which consists in positions of atomic levels and parameters of radiative transitions. The Grotrian diagram is a convenient method of representation of this information. Alongside with indicated information, these diagrams draw atomic states and their quantum numbers. The Grotrian diagrams are contained in some books (for example, [8, 15–18, 71–75]. These diagrams are most obvious for atoms and ions with simple electron shells. We give below the Grotrian diagrams for atoms with the electron shells s, s^2, and with the electron shell p^k for light atoms. These diagrams are taken from [8] and are given in Figs. 3.1, 3.3, 4.2, 4.3, 4.4, 4.5, 4.6, 4.7, 4.8, 4.9, 4.10, 4.11, 4.12, 4.13, 4.14, 4.15, 4.16, 4.17, 4.18, 4.19, 4.20, 4.21, 4.22, 4.23, 4.24, 4.25, 4.26 and 4.27. In addition, Figs. 3.6, 3.7, 3.8, 3.9, 3.10, 3.11, 3.12, 3.13, 3.15, 3.16, 3.17 and 3.18 contain positions of lowest levels of some atoms and parameters of radiative transitions between them [19]. The most information for the rates of radiative transitions between atom states is collected by NIST information center [77–81]. In addition, we use notations of LS and jj coupling schemes for excited states of gas atoms with valence p-electrons given in Table 3.10. Along with this, Pashen notations are added in Fig. 3.14 for excited inert gas atoms, as well as it is given in Figs. 3.15, 3.16, 3.17 and 3.18.

Excitation energies of the lowest excited atomic states are given in Fig. 3.22 and together with parameters of radiative dipole transitions from resonantly excited states into the ground atomic state. Energetic and radiative parameters of these states are represented in Fig. 4.1. On contrary, lower atomic states with a large lifetime are named as metastable states. The latter means that dipole radiative transitions from these states in lower ones are forbidden. In excited gases or plasmas the lifetime of metastable atoms is determined by collision processes, and they may be gathered in these systems up to more high densities than resonantly excited ones,. Therefore metastable states can influence on properties of such systems and processes in them. In addition to Fig. 3.22, Table 4.4 contains parameters of metastable atoms.

4.3 Radiation of Lowest Excited States of Inert Gas Atoms

We consider in Sect. 3.4 excited states of inert gas atoms with the electron shell $np^5(n+1)s$, and now will be analyzed the correlation between resonantly excited states within the framework of the scheme (3.23). In the case of validity of LS-coupling scheme only the state $^1P_1$ is the resonantly excited one, but in the real case depending on relation between parameters a (spin-orbit interaction) and b (exchange interaction) states $^3P_1$ and $^3P_1$ are mixed, and then the radiation processes take place from both these states. The wave functions Ψ_2 and Ψ_4 of these states are expressed through nonperturbed wave functions Φ_2 and Φ_4 in the following manner

$$\Psi_2 = c_2\Phi_2 + c_4\Phi_4, \quad \Psi_4 = -c_4\Phi_2 + c_2\Phi_4, \tag{4.13}$$

where the amplitude are given by

$$c^2_{2,4} = \frac{\sqrt{1+x^2} \pm x}{2\sqrt{1+x^2}}, \quad x = \frac{1}{2\sqrt{2}}\left(1 - \frac{3b}{\Delta_f}\right) \qquad (4.14)$$

Correspondingly, the ratio of radiative lifetimes $\tau(1s_2)$ and $\tau(1s_4)$ of these states are [61]

$$\frac{\tau(1s_2)}{\tau(1s_4)} = \frac{c^2_2}{c^2_4} = \frac{\sqrt{1+x^2} - |x|}{\sqrt{1+x^2} + |x|} \qquad (4.15)$$

Table 4.5 contains the radiative lifetimes for resonantly excited states of inert gas atoms and their ratio according to the model (3.23) under consideration [61]. Comparison of these values with measured ones [8, 77–80] exhibits the accuracy of this model that is estimated as ~20%. This analysis shows also that the accuracy of energetic parameters of excited states is better than those for radiative atomic parameters.

Thus, the analysis of energetic levels for the lowest excited shell of inert gas atoms and their radiative parameters demonstrates general positions of the atom shell model. This model, along with the interaction of electrons with the Coulomb center and Coulomb interaction between electrons, includes also exchange interaction between electrons due to the Pauli exclusion principle and spin-orbit interaction for valence electrons. In the limiting cases, depending on the ratio between two last interactions, the LS or jj coupling schemes of momenta are realized. The shell atom model neglects the correlation between electrons which occurs due to violation of the one-electron approach. The correlation between electrons is not significant for energetic atomic parameters, but it is more essential for radiative atomic parameters, and becomes of importance for two-electron and many-electron transitions in atomic particles.

4.4 Absorption Parameters and Broadening of Spectral Lines

The character of absorption and emission of photons in a gas or plasma depends on properties of this system. We consider these processes when they are determined by transition between atom discrete states. Then absorption and emission of resonant photons which energy is nearby to the difference of atom state energies, is determined by broadening of spectral lines [57, 82], i.e. broadening of energies of atom states which partake in the radiative process. Let us introduce the distribution function a_ω over frequencies of emitting photons, so that $a_\omega d\omega$ is the probability that the photon frequency ranges from ω up to $\omega + d\omega$. As the probability, the frequency distribution function of photons is normalized as

$$\int a_\omega d\omega = 1 \tag{4.16}$$

Because of spectral lines are narrow, the distribution function of photons in scales of frequencies of emitting photons is given by

$$a_\omega = \delta(\omega - \omega_o) , \tag{4.17}$$

where ω_o is the frequency of an emitting photon. Hence, the rate and times of spontaneous radiative transitions are independent of the width and shape of a spectral line, i.e. of the photon distribution function a_ω. But it is not so when photons are generated or absorbed as a result of interaction between the radiation field and atoms. There are such transitions of two types, absorption of photons and induced emission of photons. The first process is described by the scheme

$$\hbar\omega + A \rightarrow A^*, \tag{4.18}$$

and the process of induced emission is

$$n\hbar\omega + A^* \rightarrow (n+1)\hbar\omega + A, \tag{4.19}$$

where A, A^* denote an atom in he lower and upper states of transition, $\hbar\omega$ is the photon, and n is a number of identical incident photons. The absorption cross section σ_{abs} as a result of transition between two atom states is given by [57, 65]

$$\sigma_{abs} = \frac{\pi^2 c^2}{\omega^2} A a_\omega = \frac{\pi^2 c^2}{\omega^2} \frac{g_*}{g_o} \frac{a_\omega}{\tau} \tag{4.20}$$

In the same manner we have the following formula for the cross section of stimulated photon emission σ_{em} [57, 65]

$$\sigma_{em} = \frac{\pi^2 c^2}{\omega^2} B a_\omega = \frac{\pi^2 c^2}{\omega^2} \frac{a_\omega}{\tau} = \frac{g_o}{g_*} \sigma_{abs} \tag{4.21}$$

Here A, B are the Einstein coefficients, indices o and $*$ correspond to the lower and upper atom states, g_o, g_* are the statistical weighs of these atom states, τ is the radiative lifetime of the upper atom state with respect to transition to the lower atom state.

The absorption coefficient k_ω is

$$k_\omega = N_o \sigma_{abs} - N_* \sigma_{em} = N_o \sigma_{abs} \left(1 - \frac{N_* \, g_o}{N_o \, g_*} \right) \tag{4.22}$$

where N_o, N_* are the number density of atoms in the lower and upper states of transition. The absorption coefficient k_ω for a weak radiation intensity I_ω is defined from the Lambert-beer law [83, 84]

$$\frac{dI_\omega}{dx} = -k_\omega I_\omega \, , \tag{4.23}$$

where x is the direction of radiation propagation, and according to definition, the radiation flux does not perturbed a gas where it propagates. In the case when population of atom states is determined by the Boltzmann formula, the absorption coefficient is equal to

$$k_\omega = N_o \sigma_{abs} \left[1 - \exp\left(-\frac{\hbar\omega}{T} \right) \right] \tag{4.24}$$

Thus, the character of photon absorption depends on broadening of spectral lines, and we give below three types of broadening that are of importance for excited gases. The basis of Doppler broadening is the Doppler effect, according to which an emitting frequency ω_o at a relative velocity v_x of a radiating particle toward the receiver is conceived as a frequency

$$\omega = \omega_o \left(1 + \frac{v_x}{c} \right) , \tag{4.25}$$

where c is the light velocity. Hence, for radiating atomic particles with the Maxwell distribution function on velocities the distribution function of photons a_ω has the form

$$a_\omega = \frac{1}{\omega_o} \left(\frac{mc^2}{2\pi T} \right)^{1/2} \cdot \exp\left[-\frac{mc^2(\omega - \omega_o)^2}{2T\omega_o^2} \right], \tag{4.26}$$

where ω_o is the frequency for a motionless particle, m is the particle mass, T is the gas temperature expressed in energetic units.

The Lorentz (or impact) mechanism of broadening of spectral lines results from single collisions of a radiating atom with surrounding particles. As a result of these collisions, the spectral line is shifted and broaden, and the distribution function a_ω of radiating photons has the Lorenz form

$$a_\omega = \frac{1}{2\pi\nu} \left[(\omega - \omega_o + \Delta\nu)^2 + \left(\frac{\nu}{2} \right)^2 \right], \tag{4.27}$$

Here $\nu = N v \sigma_t$, where N is the number density of perturbed particles, σ_t is the total cross section of collision of a radiating and surrounding particles for an upper state of a radiating particle under assumption that collision in the lower particle state is not important. Next, $\Delta \nu = N v \sigma^*$, and if the cross section is determined by a large number of collision momenta, we have $\sigma_t \gg \sigma^*$ and one can ignore the spectral line shift. Next, in the classical case, where the main contribution to the total cross section σ_t results from many collision momenta, it is given by

$$\sigma_t = \pi R_t^2, \quad \frac{R_t U(R_t)}{\hbar v} \sim 1 , \tag{4.28}$$

where v is the collision velocity, $U(R)$ is the difference of interaction potentials for the upper and lower states of transition, and R_t is the Weiskopf radius. The criterion of validity of the Lorenz broadening (4.27) is based on the assumption that the probability to locate for two and more surrounding particles in a region of a strong interaction with a radiating atom is small, that has the form

$$N R_t^3 \ll 1 , \tag{4.29}$$

Table 4.6 compares the Doppler $\Delta \omega_D$ and Lorenz $\Delta \omega_L$ widths for transition between the ground and resonantly excited states of alkali metal atoms where these widths are given by formulas

$$\Delta \omega_D = \omega_o \sqrt{\frac{T}{mc^2}}; \quad \Delta \omega_L = \frac{1}{2} N \overline{v \sigma_t}, \tag{4.30}$$

This Table contains also the number density N_{DL} of atoms at which these line widths for indicated mechanisms of broadening are coincided.

One more mechanism of broadening of spectral lines due to interaction with surrounding atoms takes place at large number densities of surrounding atoms and corresponds to the quasistatic theory of broadening of spectral lines. As a matter, in this case a system of interacting atoms emits radiation instead of individual atom, though this interaction is small [76]. One can assume surrounding atoms to be motionless in the course of radiation, and the shift of the spectral line of a radiating atom corresponds to a certain configuration of surrounding atoms

$$\Delta \omega \equiv \omega - \omega_o = \frac{1}{\hbar} \sum_k U(R_k) , \tag{4.31}$$

where R_k is the coordinate of $k-$th atom in the frame of reference where a test radiating atom is the origin. The photon distribution function is equal for the wing of a spectral line in the case of an uniform distribution of surrounding atoms is

$$a_\omega d\omega = N \cdot 4\pi R^2 dR; \quad a_\omega = \frac{4N\pi R^2 \hbar}{dU/dR}, \tag{4.32}$$

In the case of radiative transitions between the resonantly excited and ground atom states in a parent gas or vapor, the interaction potential depends as R^{-3} on a distance R between atoms, and the total cross section is $\sigma_t \backsim 1/v$, so that $v\sigma_t$ is independent of the collision velocity. Hence, the absorption coefficient in the line center k_o is independent of the number density of atoms and their temperature. Table 4.7 contains their values for transitions between the ground and lowest resonantly excited states of atoms of alkali metal and alkali earth metal atoms. The number density N_i of alkali metal atoms given in Table 4.6 corresponds to the boundary between the Lorenz mechanism and quasistatic theory of broadening of spectral lines.

Figures

Group \ Period	I	II	III	IV	V
Resonantly excited					
1	1s ^2P (121.56) $_1$H $^2S_{1/2}$ [0.416] {1.6} Hydrogen	$1s^2$ $^1P_1$(58.433) $^1S_0$ [0.276] {0.56} $_2$He Helium			
2	2s $^2P_{1/2}$ (670.79) [0.247]{27} $_3$Li $^2S_{1/2}$ $^2P_{3/2}$(670.78) [0.494]{28} Lithium	$2s^2$ $^1P_1$(234.86) [1.34]{1.9} $_4$Be $^1S_0$ Beryllium	$^2S_{1/2}$(249.68) [0.12]{3.6} $^2P_{1/2}$ 2p $^2D_{3/2}$(208.89) [0.046]{20} $_5$B Boron	$^3D_1$ (156.03) [0.1]{8.0} $2p^2$ $^3P_0$ C $_6$ Carbon	3s ^{4}P (120) [0.27]{2.5} $2p^3$ $^4S_{3/2}$ 2s2p^4 ^{4}P (113.4) [0.08]{7.2} $_7$N Nitrogen
3	3s $P_{1/2}$ (589.59) [16]{0.318} $_{11}$Na $^2S_{1/2}$ $P_{3/2}$ (589.00) [16]{0.637} Sodium	$3s^2$ $^1P_1$ (285.21) [1.9]{2.1} $_{12}$Mg $^1S_0$ Magnesium	$^2S_{1/2}$ (394.40) [0.12]{6.8} 3p $^2P_{1/2}$ $^2D_{3/2}$ (308.22) [0.18]{13} $_{13}$Al Aluminium	$^3P_1$ (251.43) [0.17]{5.9} $3p^2$ $^3P_0$ Si $_{14}$ Silicon	$^4P_{1/2}$(178.77) [0.05]{4.0} $^4P_{3/2}$(178.28) [0.10]{4.0} $^4P_{5/2}$(177.50) [0.15]{4.0} $3p^3$ $^4S_{3/2}$ P $_{15}$ Phosphorus
4	4s $^2P_{1/2}$ (769.90) [0.35]{27} $_{19}$K $^2S_{1/2}$ $^2P_{3/2}$(766.49) [0.70]{27} Potassium	$4s^2$ $^1P_1$ (422.67) [1.7]{4.6} $_{20}$Ca $^1S_0$ Calcium	$3d4s^2$ Sc $_{21}$ $^2D_{3/2}$ Scandium	$3d^24s^2$ $_{22}$Ti $^3F_2$ Titanium	$3d^34s^2$ $_{23}$V $^4F_{3/2}$ Vanadium
	$^2P_{1/2}$ (327.40) [0.22]{7} $3d^{10}4s$ $^2S_{1/2}$ $^2P_{3/2}$(324.75) [0.44]{7.2} $_{29}$Cu Copper	$^1P_1$ (213.86) [1.5]{1.4} $^1S_0$ $4s^2$ $_{30}$Zn Zinc	$^2S_{1/2}$ (403.30) [0.12]{6.2} 4p $^2P_{1/2}$ $^2D_{3/2}$ (287.42) [0.30]{4.7} $_{31}$Ga Gallium	$4p^2$ $^3P_0$ $_{32}$Ge Germanium	$4p^3$ $^4S_{3/2}$ $_{33}$As Arsenic
5	5s $^2P_{1/2}$ (794.76) [0.32]{28} $_{37}$Rb $^2S_{1/2}$ $^2P_{3/2}$(780.03) [0.67]{26} Rubidium	$5s^2$ $^1P_1$ (460.73) [2.0]{6.2} $_{38}$Sr $^1S_0$ Strontium	$4d5s^2$ $_{39}$Y $^2D_{3/2}$ Yttrium	$4d^25s^2$ $_{40}$Zr $^3F_2$ Zirconium	$4d^45s$ $_{41}$Nb $^6D_{1/2}$ Niobium
	$^2P_{1/2}$ (338.29) [0.22]{7.9} $4d^{10}5s$ $^2S_{1/2}$ $^2P_{3/2}$ (328.07) [0.45]{6.7} $_{47}$Ag Silver	$^1P_1$ (228.80) [1.4]{1.7} $^1S_0$ $5s^2$ $_{48}$Cd Cadmium	$^2S_{1/2}$ (410.18) [0.14]{7.4} 5p $^2P_{1/2}$ $^2D_{3/2}$ (303.94) [0.36]{7.0} $_{49}$In Indium	$5p^2$ $^3P_0$ $_{50}$Sn Tin	$5p^3$ $^4S_{3/2}$ $_{51}$Sb Antimony
6	6s $^2P_{1/2}$ (894.35) [0.39]{31} $_{55}$Cs $^2S_{1/2}$ $^2P_{3/2}$ (852.11) [0.81]{28} Cesium	$6s^2$ $^1P_1$ (553.55) [1.6]{8.5} $_{56}$Ba $^1S_0$ Barium	$5d6s^2$ $_{57}$La $^2D_{3/2}$ Lanthanum	$5d^26s^2$ $_{72}$Hf $^3F_2$ Hafnium	$5d^36s^2$ $_{73}$Ta $^4F_{3/2}$ Tantalum
	$^2P_{1/2}$ (267.60) [0.12]{6.0} $5d^{10}6s$ $^2S_{1/2}$ $^2P_{3/2}$ (242.80) [0.26]{4.6} $_{79}$Au Gold	$^1P_1$ (184.95) [1.2]{1.3} $5d^{10}6s^2$ $^1S_0$ $^3P_1$ (253.65) [0.024]{120} $_{80}$Hg Mercury	$^2S_{1/2}$ (377.57) [0.13]{7.6} 6p $^2P_{1/2}$ $^2D_{3/2}$ (276.79) [0.29]{7.0} $_{81}$Tl Thallium	$^3P_1$ (283.31) [0.21]{5.8} $6p^2$ $^3P_0$ $_{82}$Pb Lead	$^4P_{1/2}$(306.77) [0.15]{4.6} $6p^3$ $^4S_{3/2}$ $_{83}$Bi Bismuth

Fig. 4.1 Resonantly excited atom states

atom states

VI	VII	VIII
^3S (130.4) 2p^4 [0.05] {1.8} $^3P_2$ ^3D (102.7) $_8$O [0.01] {25} Oxygen	$^2P_{1/2}$ (95.48) 2p^5 [0.07] {3.5} $^2P_{3/2}$ $^2P_{3/2}$ (95.19) $_9$F [0.035] {3.5} Fluorine	$^3P_1$ (74.372) $^1S_0$ 2p^6 [0.01] {25} $^1P_1$ (73.590) $_{10}$Ne [0.15] {1.6} Neon
$^3S_1$ (180.73) 3p^4 [0.11] {2.8} $^3P_2$ $_{16}$S Sulfur	$^2P_{1/2}$ (134.72) 3p^5 [0.11] {1.0} $^2P_{3/2}$ $^2P_{3/2}$ (133.57) $_{17}$Cl [0.023] {1.0} Chlorine	$^3P_1$ (106.66) 2p^6 [0.051] {20} $^1S_0$ $^1P_1$ (104.82) $_{18}$Ar [0.25] {10} Argon

3d^{5}4s $_{24}$Cr $^7S_3$ Chromium	3d^{5}4s^2 $_{25}$Mn $^6S_{5/2}$ Manganese	3d^{6}4s^2 $_{26}$Fe $^5D_4$ Iron	3d^{7}4s^2 $_{27}$Co $^4F_{9/2}$ Cobalt	3d^{8}4s^2 $_{28}$Ni $^3F_4$ Nickel
4p^4 $^3P_2$ $_{34}$Se Selenium	4p^5 $^2P_{3/2}$ $_{35}$Br Bromine	$^3P_1$ (123.58) $^1S_0$ 4p^6 [0.15] {4.4} $^1P_1$ (116.46) $_{36}$Kr [0.14] {4.5} Krypton		
4d^{5}5s $_{42}$Mo $^7S_3$ Molibdenum	4d^{5}5s^2 $_{43}$Tc $^6S_{5/2}$ Technetium	4d^{7}5s $_{44}$Ru $^5F_5$ Ruthenium	4d^{8}5s $_{45}$Rh $^4F_{9/2}$ Rhodium	4d^{10} $_{46}$Pd $^1S_0$ Palladium
5p^4 $^3P_2$ $_{52}$Te Tellurium	5p^5 $^2P_{3/2}$ $_{53}$I Iodine	$^3P_1$ (146.96) $^1S_0$ 5p^6 [0.27] {3.6} $^1P_1$ (129.56) $_{54}$Xe [0.22] {3.5} Xenon		
5d^{4}6s^2 $_{74}$W $^5D_0$ Tungsten	5d^{5}6s^2 $_{75}$Re $^6S_{5/2}$ Rhenium	5d^{6}6s^2 $_{76}$Os $^5D_4$ Osmium	5d^{7}6s^2 $_{77}$Ir $^4F_{9/2}$ Iridium	5d^{9}6s $_{78}$Pt $^3D_3$ Platinum
6p^4 $^3P_2$ $_{84}$Po Polonium	6p^5 $^2P_{3/2}$ $_{85}$At Astatine	6p^6 $^1S_0$ $_{86}$Rn Radon		

Legend:

Wavelength of radiative transition, nm — Shell of valence electrons

Electron terms of excited states — Electron term of the ground state

Oscillator strength for radiative transition — Symbol — Atomic number — Element

Lifetime of resonantly excited state, ns

$^1P_1$ (184.95)   5d^{10}6s^2
 [1.2] {1.3} $^1S_0$
 $^3P_1$ (253.65) $_{80}$Hg
 [0.024] {120}   Mercury

Fig. 4.1 (continued)

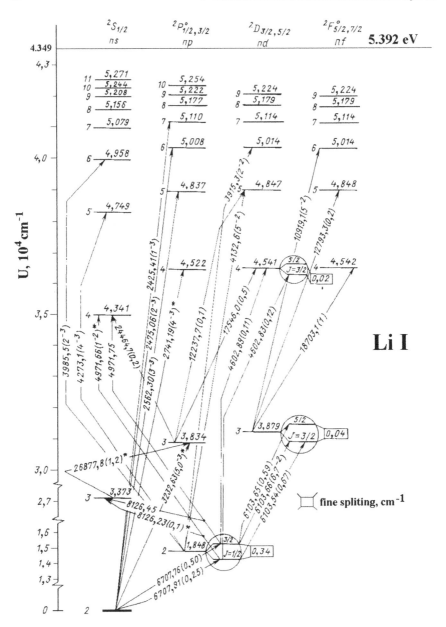

Fig. 4.2 Spectrum of lithium atom

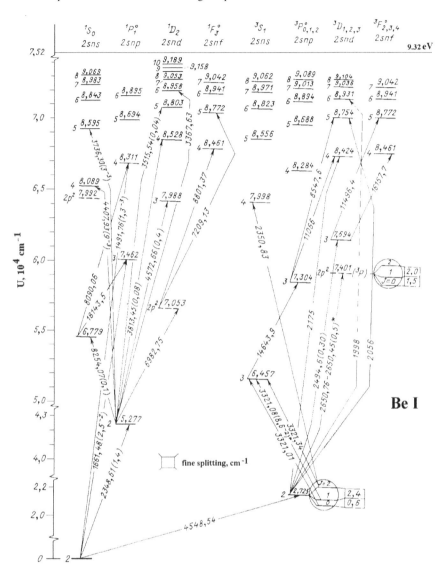

Fig. 4.3 Spectrum of beryllium atom

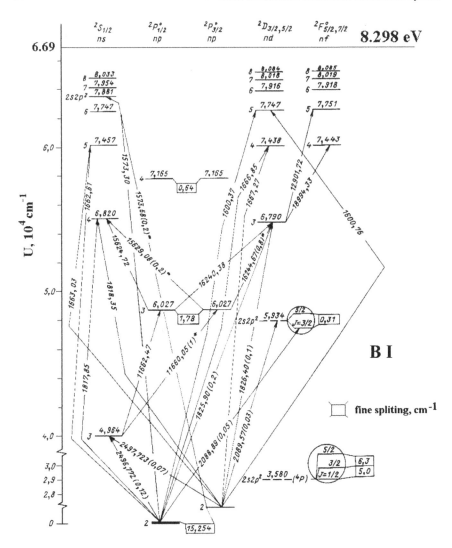

Fig. 4.4 Spectrum of boron atom

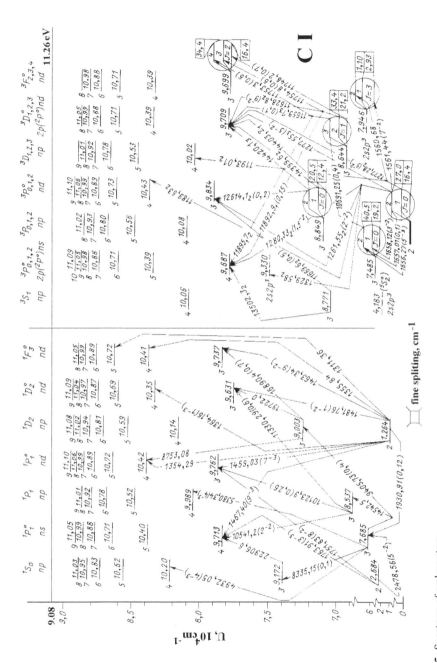

Fig. 4.5 Spectrum of carbon atom

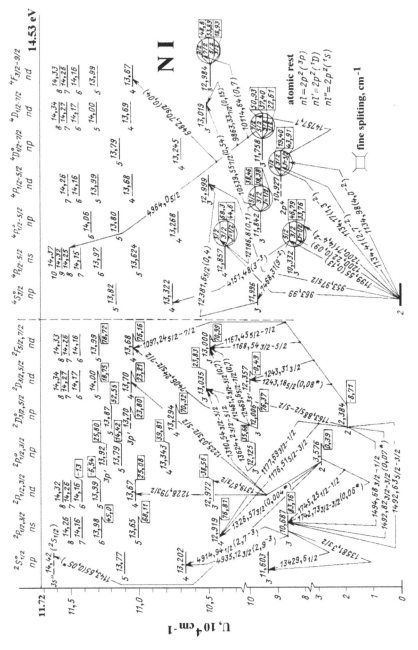

Fig. 4.6 Spectrum of nitrogen atom

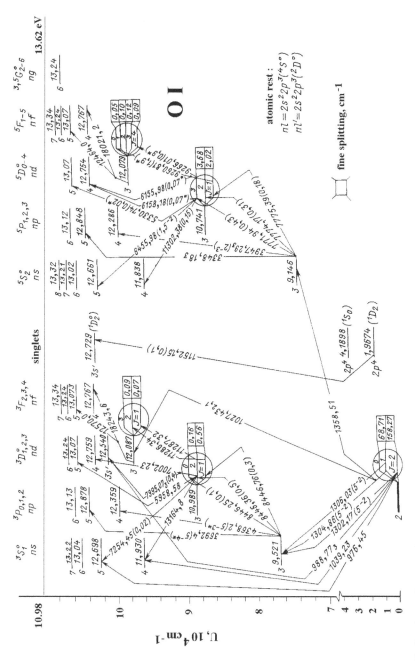

Fig. 4.7 Spectrum of oxygen atom

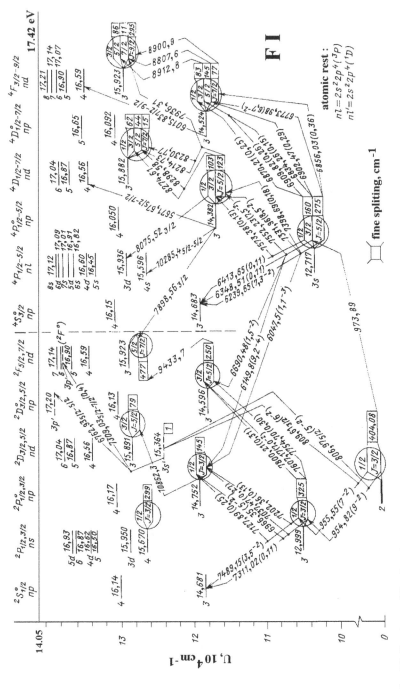

Fig. 4.8 Spectrum of fluorine atom

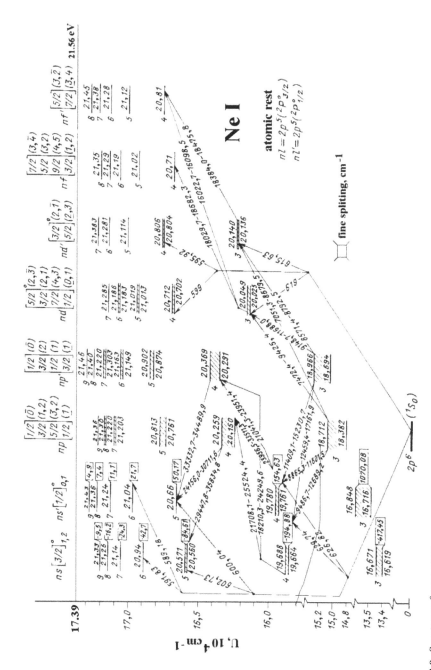

Fig. 4.9 Spectrum of neon atom

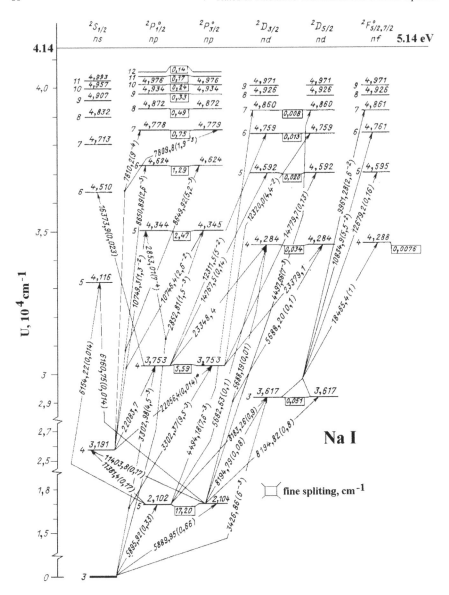

Fig. 4.10 Spectrum of sodium atom

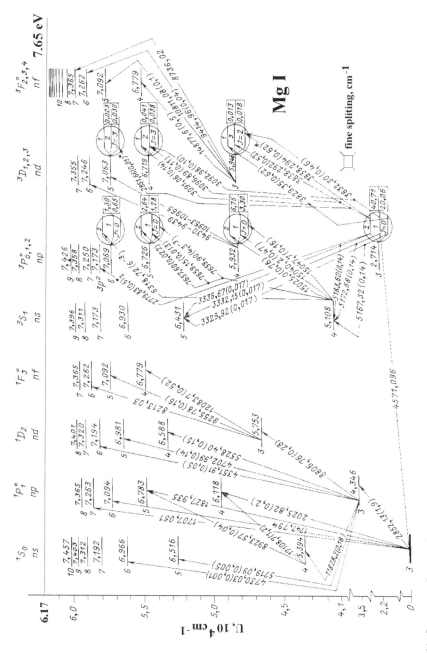

Fig. 4.11 Spectrum of magnesium atom

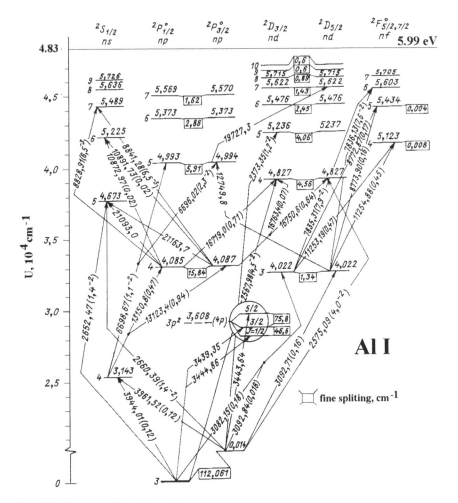

Fig. 4.12 Spectrum of aluminium atom

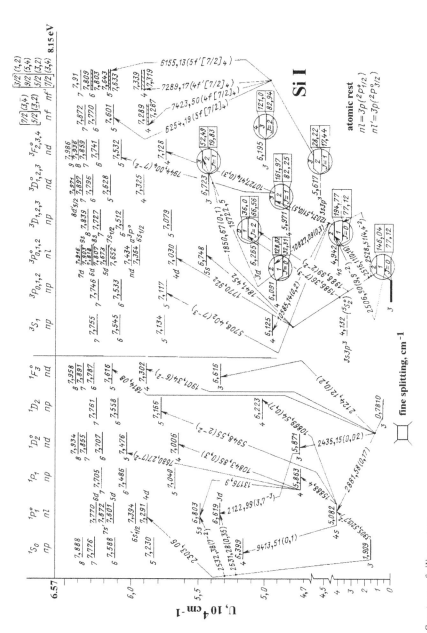

Fig. 4.13 Spectrum of silicon atom

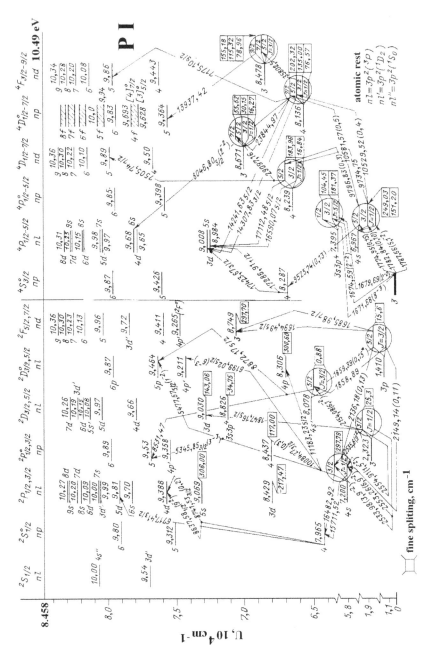

Fig. 4.14 Spectrum of phosphorusatom

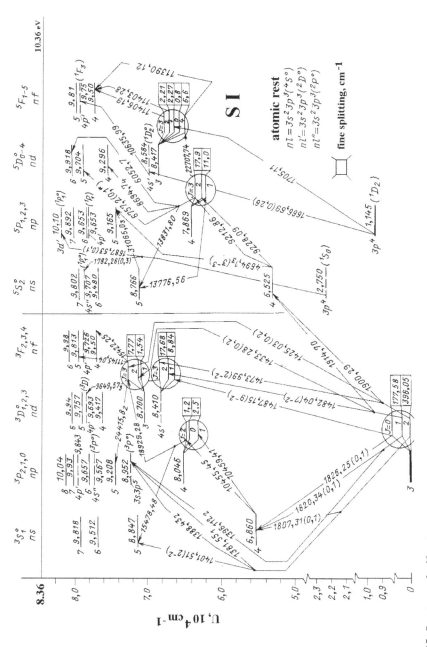

Fig. 4.15 Spectrum of sulfur atom

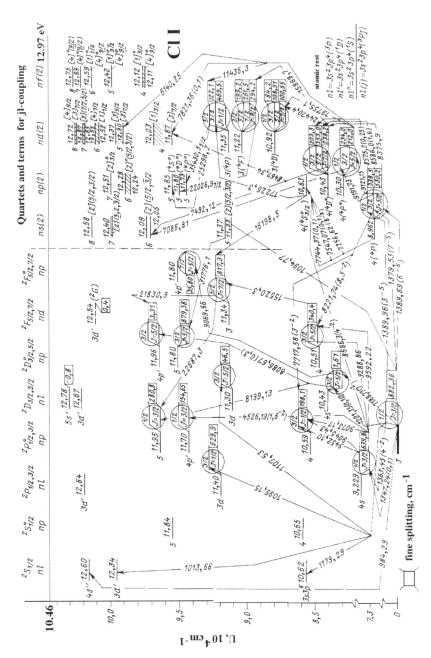

Fig. 4.16 Spectrum of chlorine atom

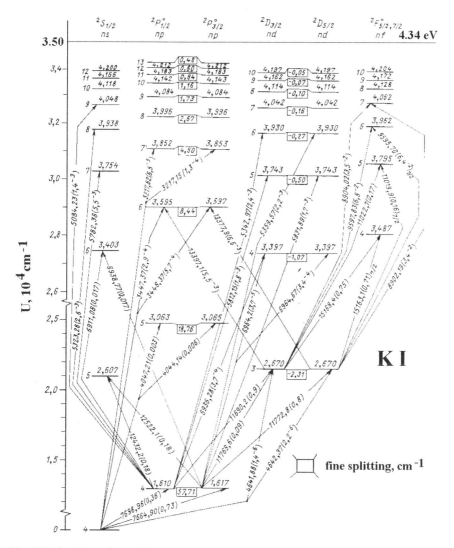

Fig. 4.17 Spectrum of potassium atom

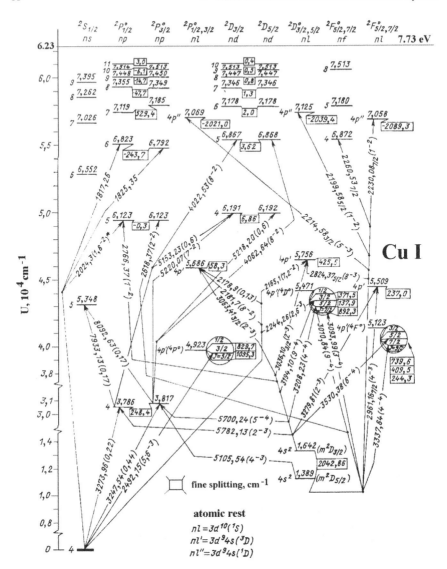

Fig. 4.18 Spectrum of copper atom

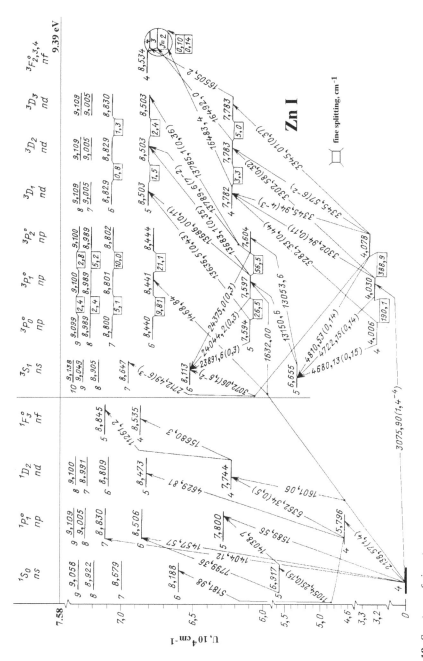

Fig. 4.19 Spectrum of zinc atom

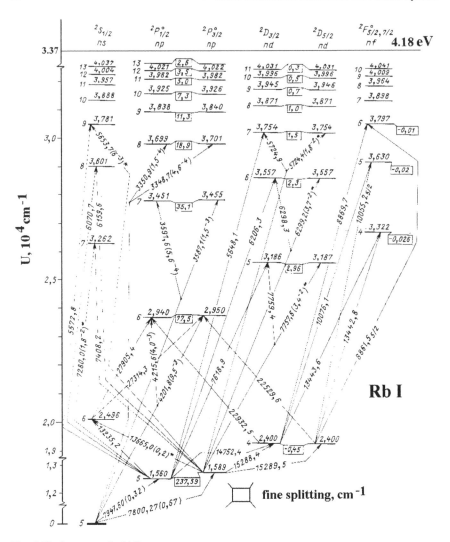

Fig. 4.20 Spectrum of rubidium atom

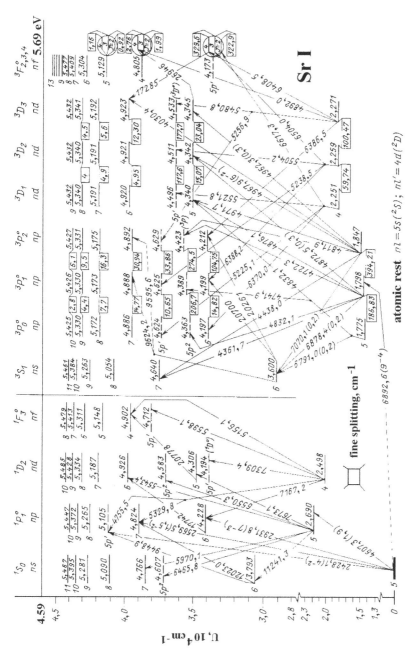

Fig. 4.21   Spectrum of strontium atom

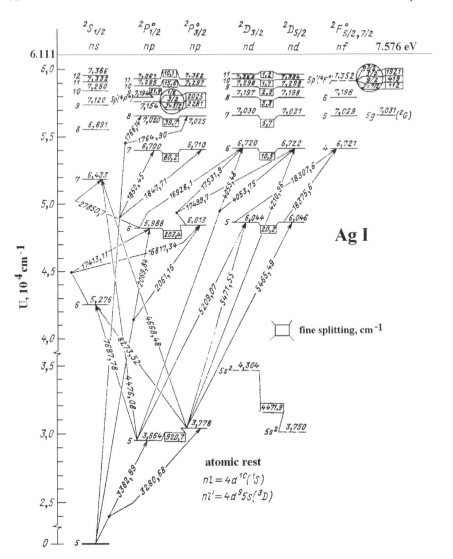

Fig. 4.22 Spectrum of silver atom

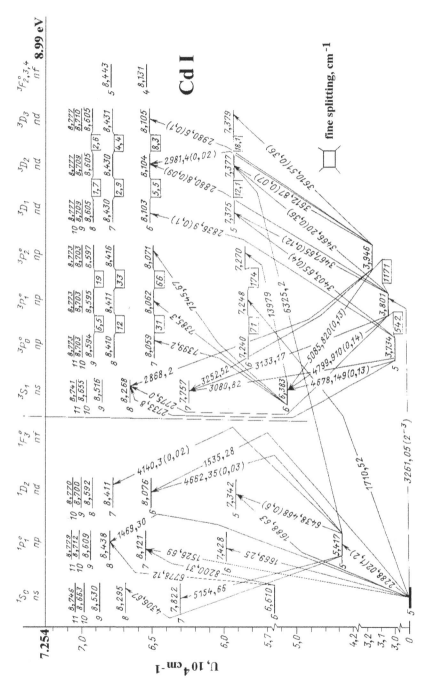

Fig. 4.23 Spectrum of cadmium atom

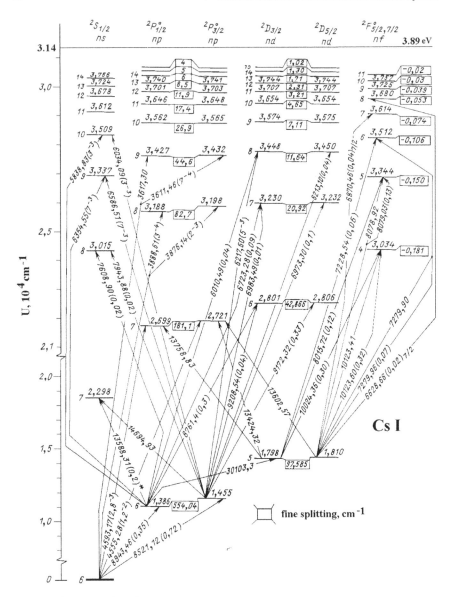

Fig. 4.24 Spectrum of caesium atom

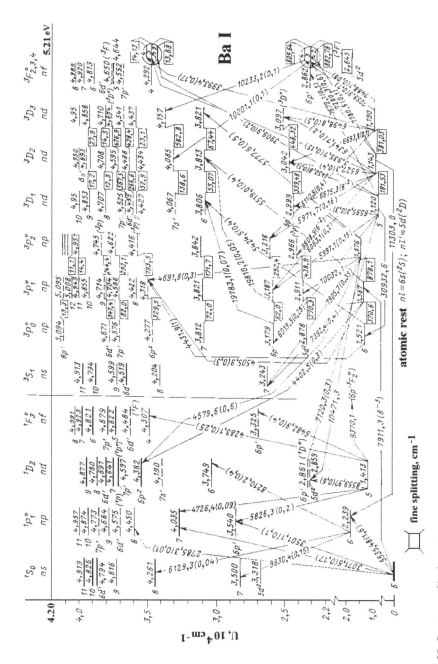

Fig. 4.25 Spectrum of barium atom

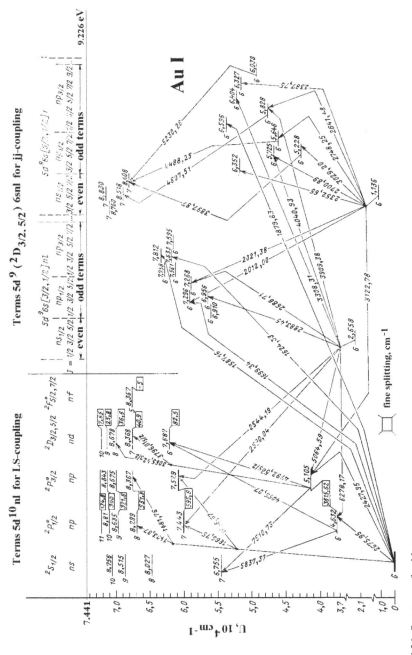

Fig. 4.26 Spectrum of gold atom

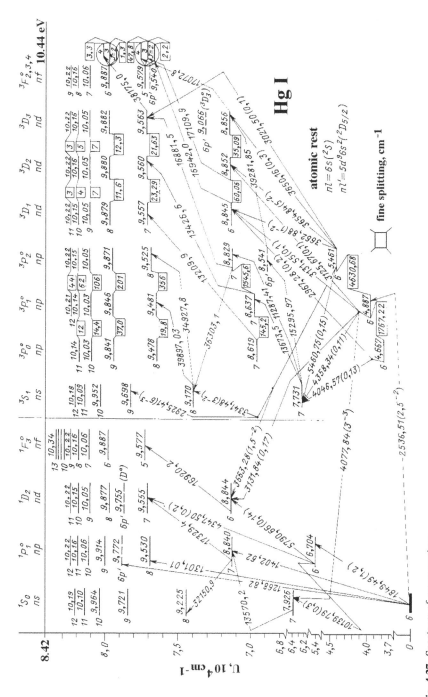

Fig. 4.27 Spectrum of mercury atom

Tables

Table 4.1 The conversional factors for radiative transition between atom states

Number	Formula	Conversional factor C	Used units
1.	$\varepsilon = C\omega$	$C = 4.1347 \cdot 10^{-15}$ eV	ω in s^{-1}
		$C = 6.6261 \cdot 10^{-34}$ J	ω in s^{-1}
2.	$\omega = C\varepsilon$	$C = 1.519 \cdot 10^{15}$ s^{-1}	ε in eV
		$C = 1.309 \cdot 10^{11}$ s^{-1}	ε in K
3.	$\omega = C/\lambda$	$C = 1.884 \cdot 10^{15}$ s^{-1}	ε in eV
4.	$\varepsilon = C/\lambda$	1.2398 eV	λ in μm
5.	$f_{o*} = C\omega d^2 g_*$	$1.6126 \cdot 10^{-17}$	ω in s^{-1}, d in D^*
		0.02450	$\Delta\varepsilon = \hbar\omega$ in eV , d in D^a
6.	$f_{o*} = Cd^2 g_*/\lambda$	0.03038	λ in μm, d in D^a
7.	$1/\tau_{*o} = C\omega^3 d^2 g_o$	$3.0316 \cdot 10^{-40}$ s^{-1}	ω, s^{-1}, d, D^a
		$1.06312 \cdot 10^6$ s^{-1}	$\Delta\varepsilon = \hbar\omega$ in eV, d in D^a
8.	$1/\tau_{*o} = Cd^2 g_o/\lambda^3$	$2.0261 \cdot 10^6$ s^{-1}	λ in μm, d in D^a
9.	$1/\tau_{*o} = C\omega^2 g_o f_{o*}/g_*$	$1.8799 \cdot 10^{-23}$ s^{-1}	ω in s^{-1}, d in D^a
		$4.3393 \cdot 10^7$ s^{-1}	$\Delta\varepsilon = \hbar\omega$ in eV; d in D^a
10.	$1/\tau_{*o} = Cf_{o*}g_o/(g_*\lambda^2)$	$6.6703 \cdot 10^7$ s^{-1}	λ in μm, d in D^a

$^a D$ is Debye, $1\ D = ea_o = 2.5418 \cdot 10^{-18}\ CGSE$

Table 4.2 Parameters of radiative transitions involving lowest states of the hydrogen atom, so that i is the lower state, j is the upper transition state [44, 45]

	f_{ij}	τ_{ji}, ns		f_{ij}	τ_{ji}, ns
$1s-2p$	0.4162	1.6	$3p-4s$	0.032	230
$1s-3p$	0.0791	5.4	$3p-4d$	0.619	36.5
$1s-4p$	0.0290	12.4	$3p-5s$	0.007	360
$1s-5p$	0.0139	24	$3p-5d$	0.139	70
$2s-3p$	0.4349	5.4	$3d-4p$	0.011	12.4
$2s-4p$	0.1028	12.4	$3d-4f$	1.016	73
$2s-5p$	0.0419	24	$3d-5p$	0.0022	24
$2p-3s$	0.014	160	$3d-5f$	0.156	140
$2p-3d$	0.696	15.6	$4s-5p$	0.545	24
$2p-4s$	0.0031	230	$4p-5s$	0.053	360
$2p-4d$	0.122	26.5	$4p-5d$	0.610	70
$2p-5s$	0.0012	360	$4d-5p$	0.028	24
$2p-5d$	0.044	70	$4d-5f$	0.890	140
$3s-4p$	0.484	12.4	$4f-5d$	0.009	70
$3s-5p$	0.121	24	$4f-5g$	1.345	240

Table 4.3 Times of radiative transitions between lowest states for heliumlike ions

	Z	$\tau(2^1P \to 1^1S)$, s	$\tau(2^3P \to 1^1S)$, s	$\tau(2^1S \to 1^1S)$, s	$\tau(2^3S \to 1^1S)$, s
Li$^+$	3	$3.9 \cdot 10^{-11}$	$5.6 \cdot 10^{-5}$	$5.1 \cdot 10^{-4}$	49
Ne^{+8}	10	$1.1 \cdot 10^{-13}$	$1.8 \cdot 10^{-10}$	$1.0 \cdot 10^{-7}$	$9.2 \cdot 10^{-5}$
Ca^{+18}	20	$6.0 \cdot 10^{-15}$	$2.1 \cdot 10^{-13}$	$1.2 \cdot 10^{-9}$	$7.0 \cdot 10^{-8}$
Zn^{+28}	30	$1.3 \cdot 10^{-15}$	$8.1 \cdot 10^{-15}$	$1.0 \cdot 10^{-10}$	$1.1 \cdot 10^{-9}$
Zr^{+38}	40	$5 \cdot 10^{-16}$	$1.4 \cdot 10^{-15}$	$2 \cdot 10^{-11}$	$6 \cdot 10^{-11}$
Sn^{+48}	50	$2 \cdot 10^{-16}$	$5 \cdot 10^{-16}$	$4 \cdot 10^{-12}$	$6 \cdot 10^{-12}$
$-d\ln\tau/d\ln Z$	–	4.1	4.6	7.2	10

Table 4.4 Parameters of metastable states of atoms; ε_{ex} is the excitation energy of the state, λ is the wave length of radiative transition to the ground state, τ is the radiative lifetime of the metastable state

Atom, state	ε_{ex}, eV	λ, nm	τ, s
H($2^2S_{1/2}$)	10.20	121.6	0.12
He(2^3S_1)	19.82	62.56	7900
He(2^1S_0)	20.62	60.14	0.02
C(2^1D_2)	1.26	983.7	3200
C(2^1S_0)	2.68	462.4	2
N($2^2D_{5/2}$)	2.38	520.03	$1.4 \cdot 10^5$
N($2^2D_{3/2}$)	2.38	519.8	$6.1 \cdot 10^4$
N($2^2P_{1/2,3/2}$)	3.58	1040	12
O(2^1D_2)	1.97	633.1	100
O(2^1S_0)	4.19	557.7	0.76
F($2^2P_{1/2}$)	0.050	24700	660
P($3^2D_{3/2}$)	1.41	880.0	$3 \cdot 10^3$
P($3^2D_{5/2}$)	1.41	878.6	$5 \cdot 10^3$
P($3^2P_{1/2,3/2}$)	2.32	1360	4
S(3^1D_2)	1.15	1106	28
S(3^1S_0)	2.75	772.4	0.5
Cl($3^2P_{1/2}$)	0.109	11100	80
Se(4^1D_2)	1.19	1160	1.4
Se(4^1S_0)	2.78	464.0	0.1
Br($4^1P_{1/2}$)	0.46	2713	0.9
Te(5^1D_2)	1.31	1180	0.28
Te(5^1S_0)	2.88	474.0	0.025
I($5^2P_{1/2}$)	0.94	1315	0.14
Hg(6^3P_0)	4.67	265.6	1.4

Table 4.5 Radiative lifetimes for lower resonantly excited states of inert gas atoms

Atom	Ne	Ar	Kr	Xe
$\tau(1s_2)$, ns	1.6	2.0	3.2	3.5
$\tau(1s_4)$, ns	25	10	3.5	3.6
$\tau(1s_2)/\tau(1s_4)$	16	5	1.1	1.0
c_2^2/c_4^2	13	3.8	1.1	1.4

Table 4.6 Broadening parameters for spectral lines of alkali metal atoms: λ is the photon wavelength for resonant transition, τ is the radiative lifetime of the resonantly excited atoms, $\Delta\omega_D$, $\Delta\omega_L$ are given by formula (4.30), N_{DL} is expressed in 10^{16} cm^{-3}, and N_t is given in 10^{18} cm^{-3}. The temperature of alkali metal atoms is 500 K [35]

Element	Transition	λ, nm	τ, ns	$\Delta\omega_D$, 10^9 s^{-1}	$\Delta\omega_L/N$, 10^{-7} cm^3/s	N_{DL}	N_t
Li	$2^2S \to 3^2P$	670.8	27	8.2	2.6	16	3.2
Na	$3^2S_{1/2} \to 3^2P_{1/2}$	589.59	16	4.5	1.6	15	2.5
Na	$3^2S_{1/2} \to 3^2P_{3/2}$	589.0	16	4.5	2.4	9.4	1.4
K	$4^2S_{1/2} \to 4^2P_{1/2}$	769.0	27	2.7	2.0	6.5	1.2
K	$4^2S_{1/2} \to 4^2P_{3/2}$	766.49	27	2.7	3.2	4.2	0.6
Rb	$5^2S_{1/2} \to 5^2P_{1/2}$	794.76	28	1.7	2.0	4.5	0.7
Rb	$5^2S_{1/2} \to 5^2P_{3/2}$	780.03	26	1.8	3.1	2.9	0.4
Cs	$6^2S_{1/2} \to 6^2P_{1/2}$	894.35	31	1.2	2.6	2.3	0.3
Cs	$6^2S_{1/2} \to 6^2P_{3/2}$	852.11	27	1.3	4.0	1.6	0.2

Table 4.7 Radiative parameters and the absorption coefficient for the center line of resonant radiative transitions in atoms of the first and second groups of the periodic system of elements [33, 35]

Element	Transition	λ, nm	τ, ns	g_*/g_o	$v\sigma_t$, 10^{-7} cm^3/s	k_o, 10^5 cm^{-1}
H	$1^2S \to 2^2P$	121.57	1.60	3	0.516	8.6
He	$1^1S \to 2^1P$	58.433	0.56	3	0.164	18
Li	$2^2S \to 3^2P$	670.8	27	3	5.1	1.6
Be	$2^1S \to 2^1P$	234.86	1.9	3	9.6	1.4
Na	$3^2S_{1/2} \to 3^2P_{1/2}$	589.59	16	1	3.1	1.1
Na	$3^2S_{1/2} \to 3^2P_{3/2}$	589.0	16	2	4.8	1.4
Mg	$3^1S \to 3^1P$	285.21	2.1	3	5.5	3.4
K	$4^2S_{1/2} \to 4^2P_{1/2}$	769.0	27	1	4.1	0.85
K	$4^2S_{1/2} \to 4^2P_{3/2}$	766.49	27	2	6.3	1.1
Ca	$4^1S \to 4^1P$	422.67	4.6	3	7.3	2.5
Cu	$4^2S_{1/2} \to 4^2P_{1/2}$	327.40	7.0	1	1.1	2.1
Cu	$4^2S_{1/2} \to 4^2P_{3/2}$	324.75	7.2	2	1.7	2.8
Zn	$4^1S \to 4^1P$	213.86	1.4	3	3.3	4.8
Rb	$5^2S_{1/2} \to 5^2P_{1/2}$	794.76	28	1	3.9	0.91
Rb	$5^2S_{1/2} \to 5^2P_{3/2}$	780.03	26	2	6.2	1.2
Sr	$5^1S \to 5^1P$	460.73	6.2	3	9.4	1.7
Ag	$5^2S_{1/2} \to 5^2P_{1/2}$	338.29	7.9	1	1.1	2.0
Ag	$5^2S_{1/2} \to 5^2P_{3/2}$	328.07	6.7	2	1.7	2.9
Cd	$5^1S \to 5^1P$	228.80	1.7	3	3.3	4.5
Cs	$6^2S_{1/2} \to 6^2P_{1/2}$	894.35	31	1	5.3	0.77
Cs	$6^2S_{1/2} \to 6^2P_{3/2}$	852.11	27	2	8.1	1.0
Ba	$6^1S \to 6^1P$	553.55	8.5	3	9.0	1.9
Au	$6^2S_{1/2} \to 6^2P_{1/2}$	267.60	6.0	1	0.49	1.9
Au	$6^2S_{1/2} \to 6^2P_{3/2}$	242.80	4.6	2	0.75	4.2
Hg	$6^1S \to 6^1P$	184.95	1.3	3	2.2	5.7

Chapter 5
Physics of Molecules

Abstract Various types of interaction of atomic particles at large distances between them are considered and include long-range interaction, exchange interaction, and electron interaction inside atomic particles. The correlation is analyzed between quantum numbers of atoms and diatmic molecule consisting of these atoms. Hund cases of coupling of momenta in a diatomic molecule are represented. Potential curves of diatomic molecules in lower electron states are represented. Numerical values of various molecular parameters for diatomic and polyatomic molecules are given.

5.1 Interaction Potential of Atomic Particles at Large Separations

We analyze below a long-range interaction involving two atoms or an atom with ion at large distances between them that allows one to separate different interaction types and to find the condition of formation of chemical bonds. A long-range interaction of atomic particles is determined by distribution of valence electrons, and the operator of atom interaction V results from Coulomb interaction of valence electrons with electrons of another atom and its core

$$\widehat{V} = -\sum_i \frac{e^2}{|\mathbf{r}_i - \mathbf{R}|} - \sum_k \frac{e^2}{|\mathbf{r}_k + \mathbf{R}|} + \sum_{i,k} \frac{e^2}{|\mathbf{r}_k + \mathbf{R} - \mathbf{r}_i|} + \frac{Z_1 Z_2 e^2}{R} \qquad (5.1)$$

Here $\mathbf{R}$ is the vector along the line joined nuclei, the coordinates of valence electrons $\mathbf{r}_i$, $\mathbf{r}_i$ are counted off from their nuclei (see Fig. 5.1). Expanding the operator (5.1) over a small parameter r/R, after an average over the electron distribution we obtain different types of a long-range interaction between atomic particles. The strongest term is proportional to R^{-2} and is the ion interaction with the atom dipole moment if it is not zero, as it takes place for ion interaction with an excited hydrogen atom which state is characterized by parabolic quantum numbers n, n_1, n_2, m. Then the interaction potential has the form [45, 46]

© Springer International Publishing AG, part of Springer Nature 2018   103
B. M. Smirnov, *Atomic Particles and Atom Systems*, Springer Series on Atomic,
Optical, and Plasma Physics 51, https://doi.org/10.1007/978-3-319-75405-5_5

$$U(R) = \frac{3Z\hbar^2(n_2 - n_1)}{2m_e R^2} \qquad (5.2)$$

where Z is the ion charge expressed in electron charges.

The next expansion term is proportional to R^{-3} and is interaction of the ion charge with the atom quadrupole moment

$$U(R) = \frac{Ze^2 Q}{R^3}, \qquad (5.3)$$

where Q is the component of the quadrupole moment tensor on the axis that joins nuclei. In the one-electron approximation this quantity is [85, 86]

$$Q = 2\sum_i \overline{r_i^2} P_2(\cos\theta_i) = 2\sum_i \frac{l_i(l_i + 1) - 3m_i^2}{(2l_i - 1)(2l_i + 3)} \overline{r_i^2}, \qquad (5.4)$$

where r_i, θ_i is the spherical coordinates of the valence electron. The quadrupole moment is zero for a completed electron shell, and for an average over the momentum projections. The values of the quadrupole moment for atoms with a filling p-shell are given in Table 5.1 [19, 61].

The next expansion term for the long-range interaction potential is interaction of a charge with an induced atomic dipole moment and for a unit ion charge is given by

$$U(R) = -\frac{\alpha e^2}{2R^4} \qquad (5.5)$$

where α is the atom polarizability which values are given in diagram of Fig. 5.2. The interaction potential of two atoms due to induced dipole moments (van der Waals interaction) where one of these atoms has a completed shell is equal

$$U(R) = -\frac{C_6}{R^6}, \qquad (5.6)$$

and the values of C_6 for two inert gas atoms are given in Table 5.2.

Exchange interaction of atomic particles results from overlapping of wave functions of valence electrons and is determined by a coordinate regions as it is shown in Fig. 5.3. Because long-range and exchange interactions between atomic particles are given by different coordinate regions at large distances between them, the total interaction potential is the sum of these interaction potentials. Next, according to the nature of the exchange interaction potential $\Delta(R)$, we have the following estimation in the case of exchange by one electron located in the field of identical atomic cores

$$\Delta(R) \sim |\psi(R/2)|^2 \sim \exp(-\gamma R)$$

Here $\psi(r)$ is the wave function of the valence electron, and $\gamma^2 = 2m_e J/\hbar^2$, where J is the atom ionization potential. This interaction potential with exchange by one electron corresponds to ion-atom interaction. Interaction of two atoms is accompanied by exchange by two valence electrons, and the exchange interaction potential of two atoms at large separations is $\Delta(R) \sim \exp(-2\gamma R)$. In particular, the exchange interaction potential of two hydrogen atoms is $\Delta(R) \sim \exp(-2R/a_o)$ at large distances between them.

The exchange interaction potential may have different sign that determines the possibility of formation of a chemical bond for a given electron term at approach of atomic particles. If this sign is negative, a covalence chemical bond is formed at moderate distances between atomic particles, and a number of attractive and repulsed electron terms is equal approximately. In particular, we consider interaction of a structureless ion and a parent atom with a valence s-electron. At large distances between nuclei R the eigen functions ψ_g and ψ_u of a system of interacting ion and atom correspond to the even (gerade) and odd (ungerade) states and are given by

$$\psi_g = \frac{\psi(1) + \psi(2)}{\sqrt{2}}; \quad \psi_u = \frac{\psi(1) - \psi(2)}{\sqrt{2}} \tag{5.7}$$

at large distances between nuclei. Here $\psi(1)$, $\psi(2)$ corresponds to electron location in the field of the first and second atomic core. Correspondingly, if an electron is located in the field of one core, its state is a combination of the even and odd states. In particular, the wave function for electron location in the field of the first atomic core is

$$\psi(1) = \frac{\psi_g + \psi_u}{\sqrt{2}}, \tag{5.8}$$

where

$$\psi_g = \frac{\psi(1) + \psi(2)}{\sqrt{2}}; \quad \psi_u = \frac{\psi(1) - \psi(2)}{\sqrt{2}} \tag{5.9}$$

As is seen, the atomic electron term is split in the even $\varepsilon_g(R)$ and odd $\varepsilon_u(R)$ molecular electron terms at finite distances R between nuclei, and the even term corresponds to attraction, and the odd term respects to repulsion. The exchange interaction potential $\Delta(R)$ is the difference of the energies of these states and for s-valence electron in the field of structureless cores this quantities at large R is equal to [128]

$$\Delta(R) \equiv |\,\varepsilon_g(R) - \varepsilon_u(R)\,| = A^2 R^{2/\gamma - 1} \exp(-R\gamma - 1/\gamma), \tag{5.10}$$

where A, γ are the parameters of the asymptotic expression (3.19) for the electron wave function.

Another type of exchange interaction of two atoms $A-B$ is realized if an electron term $A-B$ approaches with an electron term $A^- - B^+$, where a valence electron of one atom transfers to another atom. The Coulomb interaction of positive and negative ions leads to attraction, and atoms form ion-ion bond in this case. This bond is typical for atoms with a large affinity, in particular, for halogen atoms, and is realized into molecules consisted of halogen and alkali metal atoms. Excimer molecules have such a chemical bond.

Another type of exchange interaction takes place in the case where one atom is excited. The nature of this interaction consists in penetration of an excited electron to a non-excited atom when it is located in the field of the parent core. In the limiting case of a highly excited atom this interaction potential is given by the Fermi formula [90–92]

$$U(\mathbf{R}) = \frac{2\pi\hbar^2 L}{m_e}|\Psi(\mathbf{R})|^2 , \qquad (5.11)$$

where $\mathbf{R}$ is the non-excited atom coordinate, $\Psi(\mathbf{R})$ is the wave function of an excited electron, L is the electron scattering length for a non-excited atom. The Fermi formula (5.11) is valid if a size of electron orbit as well as a distance R between atoms exceeds significantly a size of a non-excited atom. This interaction may be used as a model one for systems and processes involving interacting atoms [93]. As is seen, there are many forms of interaction between atomic particles. At large distances between particles these interactions may be separated that allows one to ascertain the role of certain interactions.

5.2 Energetic Parameters of Diatomic Molecules

A diatomic molecule is a bound state of two atoms. The electron term of a diatomic energy is the electron energy ε of a given electron state as a function of a distance R between motionless nuclei. Figure 5.4 gives a typical dependence of the electron energy on a distance R between nuclei in the case if two atoms form a bond. A small parameter m_e/μ (m_e is the electron mass, μ is the reduced mass of molecule nuclei) is the basis of the molecule nature, and this small parameter allows us to separate the electron, vibration and rotation degrees of freedom. Indeed, a typical difference of energies for neighboring electron terms is $\varepsilon_o \sim 1\,\text{eV}$, a typical energy difference between neighboring vibration states is $\sim(m_e/\mu)^{1/2}\varepsilon_o$, and a typical energy difference for neighboring rotation states is $\sim(m_e/\mu)\varepsilon_o$.

We now give standard notations for the molecule energy [7]. The total excitation energy of the molecule T with accounting for the separation of degrees of freedom may be represented in the form [7]

$$T = T_e + G(v) + F_v(J) , \qquad (5.12)$$

where T_e is the excitation energy of an electron state, that corresponds to excitation of this electron term from the ground state of the electron term, $G(v)$ is the excitation energy of a vibrational state and $F_v(J)$ is the excitation energy of a rotational state. Formula (5.12) includes the main part of the vibrational and rotational energy for a weakly excited molecule. Because of a high resolution of spectral lines, molecular spectroscopy gives reach information about molecule energetics including parameters of electron, vibration and rotation molecular excitations [66, 69, 70, 94, 95].

Near the minimum of the electron term that corresponds to the stable molecule state, the vibrational and rotational energies of a molecule with accounting for the first expansion terms have the form

$$G(v) = \hbar\omega_e(v + 1/2) - \hbar\omega_e x(v + 1/2)^2, \quad F_v(J) = B_v J(J + 1), \quad B_v = B_e - \alpha_e(v + 1/2),$$
$$(5.13)$$

Here v and J are the vibration and rotation quantum numbers (J is the rotation momentum) which are whole numbers starting from zero. The spectroscopic parameters $\hbar\omega_e$, $\hbar\omega_e x$ are the vibration energy and the anharmonic parameter, $B_e = \hbar^2/(2I)$ is the rotation constant, where the inertia momentum equals $I = \mu R_e^2$, where R_e is the distance between atoms for the electron term minimum, B_v is the rotation constant for a vibrational excited state. These parameters for diatomic molecules and molecular ions with identical nuclei are contained in periodical tables of Figs. 5.5, 5.6, and 5.7.

5.3 Coupling Schemes in Diatomic Molecule

We below give the classification of interactions inside the diatomic molecule keeping the standard Hund scheme of momentum coupling [46, 58, 59]. Then we are based on three interaction types inside the molecule, the electrostatic interaction V_e (interaction between the orbital angular momentum of electrons and the molecular axis), spin-orbit interaction V_m and interaction between the orbital and spin electron momenta with rotation of the molecular axis V_r. A certain scheme of coupling is determined by the hierarchy of these interactions, and possible coupling scheme are given in Table 5.3.

Each type of momentum coupling gives certain quantum numbers. These quantum numbers describe the electron molecule state, correspond to a given hierarchy of interactions and are given in Table 5.3. Indeed, let us denote by $\mathbf{L}$ the electron angular momentum, by $\mathbf{S}$ the total electron spin, and by $\mathbf{j}$ the total electron momentum that is the sum of the angular and spin momenta ($\mathbf{j} = \mathbf{L} + \mathbf{S}$). Next, denote by $\mathbf{n}$ the unit vector directed along the molecular axis, and by $\mathbf{K}$ the rotation momentum of nuclei. Depending on the hierarchy of interactions, one can obtain the following quantum numbers. Indeed, Λ is the projection of the angular momentum of electrons onto the molecular axis, Ω is the projection of the total electron momentum $\mathbf{J}$ onto the molecular axis, S_n is the projection of the electron spin onto the molecular axis. L_N, S_N, J_N are projections of these momenta onto the direction of nuclei rotation momentum $\mathbf{N}$. Note that the operator of the total molecule momentum $\mathbf{J} = \mathbf{L} + \mathbf{S} + \mathbf{K} = \mathbf{j} + \mathbf{K}$ commutes with the molecule Hamiltonian, and the eigenvalues of the operator $\mathbf{J}$ are the molecule quantum numbers for various cases of Hund coupling.

The character of momentum coupling under consideration describes not only the molecule structure and quantum numbers of a diatomic molecule, but it is of importance for dynamics of atomic collisions [96–99] if transitions between states of colliding atomic particles is considered as a result of transition between different scheme of momentum coupling. Indeed, in the course of collision of two atomic particles when a distance between two colliding atomic particles varies, the character of momentum coupling may be changed also. Transition from one coupling scheme to another one leads simultaneously to a change the quantum numbers of colliding particles. This may be responsible for some transitions in atomic collisions [19, 98, 99].

The classical Hund scheme of momentum coupling inside a molecule gives a qualitative description of this problem. In reality a number of interactions is more and the hierarchy of interactions is more complex than that in the Hund cases of momentum coupling inside a molecule. We demonstrate this on a simple example of interaction between a halogen ion and its atom at large distances between them which are responsible for the charge exchange process that proceeds in collision of these atomic particles. For definiteness, we take interaction between these particle at a distance R_o, so that the cross section of resonant charge exchange is $\pi R_o^2/2$ at the collision energy of 1 eV. Because this distance is large compared to an atom or ion size, this simplifies the problem and allows us separate various types of interactions.

Let us enumerate interaction potentials in the quasimolecule X_2^+ (X is the halogen atom in the ground electron state) which are as follows

$$V_{ex}, \ U_M = \frac{Q_{MM}}{R^3}, \ U_m = \frac{Q_{MM}q_{mm}}{R^5}, \ \Delta(R), \ \delta_i, \ \delta_a, \ V_{rot} \qquad (5.14)$$

Here we divide the electrostatic interaction V_e of Table 5.3 into four parts: the exchange interaction, V_{ex}, inside the atom and ion that is responsible for electrostatic splitting of levels inside an isolated atom and ion, the long-range interaction, U_M, of the ion charge with the quadrupole moment of the atom, the long-range interaction, U_m, that is responsible for splitting of ion levels, and the ion-atom exchange interaction potential, Δ, that is determined by electron transition between atomic cores. Instead of the relativistic interaction V_m of Table 5.3 we introduce separately the fine splitting of levels δ_i for the ion (the splitting of $^3P_0$ and $^3P_2$ ion levels) and δ_a (the splitting of $^3P_{1/2}$ and $^3P_{3/2}$ atom levels) for the atom. Here M, m are the projections of the atom and ion angular momenta on the molecular axis, R is an ion-atom distance, Q_{ik} is the tensor of the atom quadrupole moment, q_{ik} is the quadrupole moment tensor for the ion. Taking for simplicity the impact parameter of ion-atom collision to be R_o, we have for the rotation energy at closest approach of colliding particles

$$V_{rot} = \frac{\hbar v}{R_o},$$

where v is the relative ion-atom velocity, and we use this expression below for estimation of the Coriolis interaction. As is seen, the number of possible coupling cases is larger than that in the classical case. Of course, only a small part of these cases can be realized.

Table 5.4 lists the values of the above interactions for the halogen molecular ion at a distance R_o between nuclei that is responsible for the resonant charge exchange process. Comparing different types of interaction, we obtain the following hierarchy of interactions in this case

$$V_{ex} >> \delta_i, \delta_a >> U_M >> U_m, V_{rot} \tag{5.15}$$

Comparing this with the data of Table 5.3, we obtain an intermediate case between cases "a" and "c" of Hund coupling, but the hierarchy sequence (5.15) does not correspond exactly to any one of the Hund cases.

Thus, we conclude that the classical Hund cases of electron momentum coupling in diatomic molecules may be used for qualitative description if the molecule structure and its quantum numbers, but in reality the interaction inside molecules is more complex.

Let us consider excimer molecules consisting of an atom in the ground state and an atom in excited state, and these atoms form a strong chemical bond. When these atoms are in the ground state, the bond is weak. Such molecules are named excimer molecules. Widespread excimer molecules contain a halogen atom in the ground state and inert gas atom in lowest excited states with an excited s-electron. These excimer molecules are analogous to molecules consisting of halogen and alkali metal atoms in the ground state, and the chemical bond in these excimer molecules result from a partial transition of an excited s-electron to the halogen atom, and interaction of positive and negative ions in these molecules determines a strong chemical bond in them.

Electron terms of the excimer molecule XeF are given in Fig. 5.8 as an example of an excimer molecule. If atoms are found in the ground states, the chemical bond between them is not realized, since an exchange interaction involving an atom with the completed electron shell corresponds to repulsion. Then valence electrons remain in the field of parent atoms, so that the quantum molecular numbers are the total molecule spin, the projection of the angular electron momentum onto the molecular axis, and also the molecule symmetry for its reflection with respect to the plane passed through the molecular axis. Excitation of the molecule leads to transition of an excited electron to the halogen atom, and the quantum number of a formed molecule is the total momentum of an inert gas atomic core. Figure 5.9 contains parameters of some states of excimer molecules under consideration [?], so that R_e is the equilibrium distance between nuclei that corresponds to the interaction potential minimum, and D_e is the potential well depth, i.e. the difference of the interaction potential at infinite distance between nuclei and the equilibrium one R_e. Radiative transitions in excimer molecules are the basis of excimer lasers [101–103], and Fig. 5.10 shows radiative transitions between electron states of excimer molecules, whereas the radiative lifetime for some states of excimer molecules as well as the wavelengths of the band middle for corresponding radiative transitions are represented in Fig. 5.9.

5.4 Potential Curves and Correlation of Atomic and Molecular States

If two atoms form a molecule, electron states of these atoms, as well as the character of coupling of electron momenta in atoms and molecules, determine possible electron states of a forming molecules. In other words, there is the correlation between atomic and molecular states [104–107]. We below consider this for a molecule consisting of two identical atoms.

If a molecule consists of identical atoms, the new symmetry occurs which corresponds to the electron reflection with respect to the plane that is perpendicular to the molecular axis and divides it in two equal parts. The electron wave function of even (gerade) state conserves its sign as a result of this operation, and the wave function of the odd (ungerade) state changes a sign when the electrons are reflected with respect to the symmetry plane. The corresponding states are denoted by g and u correspondingly.

Note also that the operator of electron reflection with respect to any plane passed through the molecular axis commutes with the electron Hamiltonian also, and the parity of a molecular state is denoted by $+$ or $-$ depending of conservation or change of the sign of the electron wave function in this operation. The parity of the molecular state is the sum of parities of atomic states, when atoms are removed on infinite distance, and this value does not change at variation of a distance between nuclei. Thus, in the "a" case of the Hund coupling molecular quantum numbers are the projection Λ of the angular momentum on the molecular axis, the total molecule spin S and its projection onto a given direction, the evenness and parity of this state which is degenerated with respect to the spin projection. Table 5.5 represents the correlation between states of two identical atoms and a forming molecule [108] in the case "a" of Hund coupling when electrostatic interaction dominates that includes the exchange interaction in atoms. This coupling character relates to molecules consisting of light atoms with the LS scheme of momentum coupling in atoms.

The "a" Hund case corresponds to light atoms for which the LS-scheme of momentum coupling is realized. Figures 5.11, 5.12, 5.13, 5.14 and 5.15 [8] give the potential curves or electron terms for lowest electron states of hydrogen, helium, carbon, nitrogen and oxygen diatomic molecules, where the "a" Hund case is realized, and the structure of valence electron shells is s, s^2, $2p^2$, $2p^3$ and $2p^4$ in these cases correspondingly. These Figures contain also electron terms of corresponding molecular ions. In these cases an electron state of an excited molecule is denoted by letters X, A, B, C etc. for the states with zero total spin and by letters a, b, c etc. for the states with one total spin. The projection of the molecule angular momentum onto the molecular axis as a quantum number is denoted by letters Σ, Π, Δ etc., when the electron momentum projection is one, two, three etc. Next, the total molecule spin is the quantum number in the "a" Hund case, and the multiplicity of the spin state $2S + 1$ (S is the molecule spin) is given as a left superscript for the state notation. Next, the state parity ($+$ or $-$) is given as a right superscript, and the state evenness for a molecule consisting of identical atoms (g or u) is represented as a right subscript.

In particular, $^3\Sigma_u^+$ describes the molecular state with the total electron spin $S = 1$, the electron momentum $\Lambda = 0$, even state u with respect to electron reflection for the plane which is perpendicular to the molecular axis and divides the molecule in two equal parts, and with the parity $+$ where the wave function conserves the sign at reflection with respect to a plane which passes through the molecular axis. The above notations are used in Figs. 5.11, 5.12, 5.13, 5.14, and 5.15.

The correlation between atomic and molecular states in the "c" case of Hund coupling, where the coupling in atoms corresponds to the jj coupling scheme, may be fulfilled in the same manner. This correlation is given in Table 5.6 that is taken from [108]. In the Hund "c" case [58, 59], when spin-orbit interaction exceeds the level splitting for different projection of the angular electron momentum, the total electron momentum, that is the molecule quantum number, is denoted by a large value. In addition, depending on the behavior of the electron wave function as a result reflection with respect to the plane passed through molecule axis, the electron state is denoted by superscripts $+$ or $-$ after the electron momentum. Transition from the Hund case "a" to "c" leads to the following change in term notations for a diatomic molecule

$$X\,^1\Sigma_g^+ \to 0_g^+ \; ; \; a\,^3\Sigma_u^+ \to 1_u\,, 0_u^-\;\; A\,^1\Sigma_u^+ \to 0_u^+ \; ; \;\; b\,^3\Sigma_g^+ \to 1_g\,, 0_g^- \qquad (5.16)$$

As an example, Fig. 5.16 contains electron terms of the argon molecule that consists of atoms in the ground the $Ar(3p^6)$ and excited $Ar(3p^54s)$ states. States of an excited atom at large separations are given for LS-scheme of momentum coupling. The same behavior of electron terms take place for molecules of other inert gases.

We note that the quantum numbers of the molecule electron state are defined accurately only for a restricted number of interactions as it takes place in the Hund cases of momentum coupling. In particular, a simple description relates to light atoms where non-relativistic interactions dominate. The description of molecules consisting of heavy atoms, when different types of interaction partakes in molecule construction, is in reality outside the Hund scheme. In particular, let us return to the case of Table 5.4 for interaction a halogen positive ion with a parent atom at large distances which are responsible for the resonant charge exchange process. In accordance with the hierarchy of interactions which is given by formula (5.15), the molecule quantum numbers are JM_Jj and the evenness (g or u), where J is the total atom momentum, M_J is its projection on the molecular axis, and j is the total ion momentum. Figure 5.16 contains electron terms of the molecular diatomic ion Cl_2^+ in a range of large distances between nuclei [100] (Fig. 5.17).

Table 5.7 contains parameters of the lowest excited states of inert gas diatomic molecules, so that R_e is the equilibrium distance between nuclei, D_e is the minimum interaction potential that corresponds to this distance (see Fig. 5.4), τ is the molecule radiative lifetime. One more peculiarity of interaction between atoms follows from the character of exchange interaction. At large separations the exchange interaction potential is given by the Fermi formula (5.11), and the electron scattering length L in this formula is positive for helium and neon atoms and is negative for argon, krypton and xenon atoms. This means that repulsion at large distances in helium and neon excited molecules is changed by attraction at moderate distances between interacting atoms. From this it follows that the interaction potential of helium or neon atoms in the ground and excited states can have the hump at large separations, and the parameters of this hump are given in Table 5.8.

In addition, Table 5.9 contains the polarizabilities of homonuclear diatomic molecules, and the diagram 5.18 gives the electron affinities of homonuclear diatomic molecules.

The properties of diatomic molecules consisting of different atoms with nearby ionization potentials are similar to those for homonuclear diatomic molecules. This is confirmed by Figs. 5.19, 5.20, and 5.21 which contain potential curves for the lowest states of molecules CH, NH, OH, and diagram Fig. 5.22, gives the affinities of various atoms to the hydrogen and oxygen atom, i.e. the dissociation energies of corresponding diatomic molecules.

5.5 Polyatomic Molecules

It will be represented below some parameters of molecules consisting of three and more atoms. Our goal in this consideration is to extract widespread information and to give a nature of these data. We first consider three-atom molecules with two or three identical atoms. These atoms may lay in one line that corresponds to the molecule symmetry $D_{\infty h}$ or may form a triangle with the symmetry C_{2v}. The geometric and vibration-rotation parameters of three-atom molecules are given in Table 5.10. This Table contains three vibrational energies ν_1, ν_2, ν_3 for these molecules, A_o, B_o, C_o are rotational constants for these molecules, r is the distance between atoms X and Y of the molecule XY_2, α is the angle between two segments XY of this molecule. In the case of the $D_{\infty h}$ symmetry, the angle $\alpha = 180°$ is and the rotational constants $A_o = C_o = 0$ because this molecule has the stick shape. In addition, as a demonstration of vibrational molecule parameters, Fig. 5.23 contains parameters of lower vibration states of the carbon dioxide molecule together with radiative parameters for these states.

Table 5.11 contains the ionization potential J and the electron affinity EA for three-atom molecules, and the binding energy D for the bond $YX - Y$ of molecules under consideration.

Table 5.12 gives the electron affinities of round fluorine-containing molecules AF_4 and AF_6 [13]. One can construct the wave function of a valence electron for these negative ions as a combination of the partial wave functions if a valence electron is located for each partial electron function at a certain fluorine atom. From this one can see existence of 4 different states of negative ions AF_4^- and six different states of negative ions in the case AF_6^- if these states are formed from atoms in the ground electron state. Correspondingly, the ground state of the negative ion is stable and is characterized by a large binding energy. In contrast to this consideration, the data of Table 5.12 show a high sensitivity of the electron binding energy in a negative ion to some details of electron interaction with a molecule. Hence, in spite of a low accuracy of data of Table 5.12, the electron binding energies in negative ions of the identical structure are significantly different.

Figures

Fig. 5.1 Geometry of interaction for bound electrons of different atoms

Polarizabilities of atoms and diatomics

Legend (key box):

Shell of valence electrons	Polarizability for atom in a_0^3 : 4.5
Atomic weight: 1.008	Symbol: $1s$ H $^2S_{1/2}$
Atomic number: 1	Polarizability for diatomic molecule in a_0^3 : 5.417
Element: Hydrogen	Electron term

Each cell below lists: atomic weight · configuration · $_Z$Symbol · electron term · element name · atom polarizability (and diatomic polarizability where given).

Group	I	II	III	IV	V	VI	VII	VIII
1	1.008 · $1s$ · $_1$H $^2S_{1/2}$ · Hydrogen · 4.5 · 5.417							4.003 · $1s^2$ · $_2$He $^1S_0$ · Helium · 1.383
2	6.941 · $2s$ · $_3$Li $^2S_{1/2}$ · Lithium · 162 · 230	9.012 · $2s^2$ · $_4$Be $^1S_0$ · Beryllium · 38	10.81 · $2p$ · $_5$B $^2P_{1/2}$ · Boron · 20.5	12.011 · $2p^2$ · $_6$C $^3P_0$ · Carbon · 11.8	14.007 · $2p^3$ · $_7$N $^4S_{3/2}$ · Nitrogen · 7.5 · 11.8	15.999 · $2p^4$ · $_8$O $^3P_2$ · Oxygen · 5.41 · 10.7	18.998 · $2p^5$ · $_9$F $^2P_{1/2}$ · Fluorine · 3.76	20.179 · $2p^6$ · $_{10}$Ne $^1S_0$ · Neon · 2.68
3	22.990 · $3s$ · $_{11}$Na $^2S_{1/2}$ · Sodium · 162 · 200	24.305 · $3s^2$ · $_{12}$Mg $^1S_0$ · Magnesium · 72	26.982 · $3p$ · $_{13}$Al $^2P_{1/2}$ · Aluminum · 160	28.086 · $3p^2$ · $_{14}$Si $^3P_0$ · Silicon · 150	30.974 · $3p^3$ · $_{15}$P $^4S_{3/2}$ · Phosphorus · 24	32.06 · $3p^4$ · $_{16}$S $^3P_2$ · Sulfur · 18	35.453 · $3p^5$ · $_{17}$Cl $^2P_{3/2}$ · Chlorine · 14 · 31	39.948 · $3p^6$ · $_{18}$Ar $^1S_0$ · Argon · 11.1
4	39.098 · $4s$ · $_{19}$K $^2S_{1/2}$ · Potassium · 290 · 410	40.08 · $4s^2$ · $_{20}$Ca $^1S_0$ · Calcium · 170	44.956 · $3d4s^2$ · $_{21}$Sc $^2D_{3/2}$ · Scandium · 59	47.88 · $3d^24s^2$ · $_{22}$Ti $^3F_2$ · Titanium	50.942 · $3d^34s^2$ · $_{23}$V $^4F_{3/2}$ · Vanadium	51.996 · $3d^54s$ · $_{24}$Cr $^7S_3$ · Chromium · 74	54.938 · $3d^54s^2$ · $_{25}$Mn $^6S_{5/2}$ · Manganese · 100	55.847 · $3d^64s^2$ · $_{26}$Fe $^5D_4$ · Iron · 90 ‖ 58.933 · $3d^74s^2$ · $_{27}$Co $^4F_{9/2}$ · Cobalt · 74 ‖ 58.69 · $3d^84s^2$ · $_{28}$Ni $^3F_4$ · Nickel · 70
4	63.546 · $3d^{10}4s$ · $_{29}$Cu $^2S_{1/2}$ · Copper · 320 · 460	65.38 · $4s^2$ · $_{30}$Zn $^1S_0$ · Zinc · 190	69.72 · $4p$ · $_{31}$Ga $^2P_{1/2}$ · Gallium · 60	72.59 · $4p^2$ · $_{32}$Ge $^3P_0$ · Germanium · 122	74.922 · $4p^3$ · $_{33}$As $^4S_{3/2}$ · Arsenic · 94	78.96 · $4p^4$ · $_{34}$Se $^3P_2$ · Selenium · 88	79.904 · $4p^5$ · $_{35}$Br $^2P_{3/2}$ · Bromine · 30 · 44	83.80 · $4p^6$ · $_{36}$Kr $^1S_0$ · Krypton · 16.7
5	85.468 · $5s$ · $_{37}$Rb $^2S_{1/2}$ · Rubidium · 320	87.62 · $5s^2$ · $_{38}$Sr $^1S_0$ · Strontium · 190	88.906 · $4d5s^2$ · $_{39}$Y $^2D_{3/2}$ · Yttrium	91.22 · $4d^25s^2$ · $_{40}$Zr $^3F_2$ · Zirconium	92.906 · $4d^45s$ · $_{41}$Nb $^6D_{1/2}$ · Niobium	95.94 · $4d^55s$ · $_{42}$Mo $^7S_3$ · Molybdenum	[98] · $4d^55s^2$ · $_{43}$Tc $^6S_{5/2}$ · Technetium	101.07 · $4d^75s$ · $_{44}$Ru $^5F_5$ · Ruthenium ‖ 102.91 · $4d^85s$ · $_{45}$Rh $^4F_{9/2}$ · Rhodium ‖ 106.42 · $4d^{10}$ · $_{46}$Pd $^1S_0$ · Palladium · 47
5	107.87 · $4d^{10}5s$ · $_{47}$Ag $^2S_{1/2}$ · Silver · 360	112.41 · $5s^2$ · $_{48}$Cd $^1S_0$ · Cadmium · 270	114.82 · $5p$ · $_{49}$In $^2P_{1/2}$ · Indium · 69	118.69 · $5p^2$ · $_{50}$Sn $^3P_0$ · Tin	121.75 · $5p^3$ · $_{51}$Sb $^4S_{3/2}$ · Antimony	127.60 · $5p^4$ · $_{52}$Te $^3P_2$ · Tellurium	126.90 · $5p^5$ · $_{53}$I $^2P_{3/2}$ · Iodine · 27	131.29 · $5p^6$ · $_{54}$Xe $^1S_0$ · Xenon · 27.4
6	132.90 · $6s$ · $_{55}$Cs $^2S_{1/2}$ · Cesium · 360	137.33 · $6s^2$ · $_{56}$Ba $^1S_0$ · Barium · 270	138.90 · $5d6s^2$ · $_{57}$La $^2D_{3/2}$ · Lanthanum	178.49 · $5d^26s^2$ · $_{72}$Hf $^3F_2$ · Hafnium	180.95 · $5d^36s^2$ · $_{73}$Ta $^4F_{3/2}$ · Tantalum	183.85 · $5d^46s^2$ · $_{74}$W $^5D_0$ · Tungsten	186.21 · $5d^56s^2$ · $_{75}$Re $^6S_{5/2}$ · Rhenium	190.2 · $5d^66s^2$ · $_{76}$Os $^5D_4$ · Osmium ‖ 192.22 · $5d^76s^2$ · $_{77}$Ir $^4F_{9/2}$ · Iridium ‖ 195.08 · $5d^96s$ · $_{78}$Pt $^3D_3$ · Platinum
6	196.97 · $5d^{10}6s$ · $_{79}$Au $^2S_{1/2}$ · Gold	200.59 · $5d^{10}6s^2$ · $_{80}$Hg $^1S_0$ · Mercury · 34	204.38 · $6p$ · $_{81}$Tl $^2P_{1/2}$ · Thallium · 46	207.2 · $6p^2$ · $_{82}$Pb $^3P_0$ · Lead	208.98 · $6p^3$ · $_{83}$Bi $^4S_{3/2}$ · Bismuth	[209] · $6p^4$ · $_{84}$Po $^3P_2$ · Polonium · 9.0	[210] · $6p^5$ · $_{85}$At $^2P_{3/2}$ · Astatine	[222] · $6p^6$ · $_{86}$Rn $^1S_0$ · Radon

Fig. 5.2 Atomic and diatomic polarizabilities

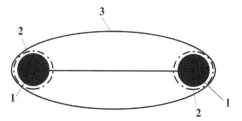

Fig. 5.3 Regions of coordinates of valence electrons which are responsible for atom interaction at large distances. 1—region of location of atomic electrons; 2—region that is responsible for a long-range atom interaction at large separations; 3—region of electron coordinates that is responsible for exchange interaction of atoms

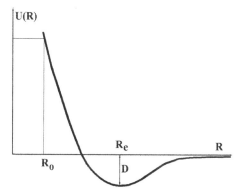

Fig. 5.4 Typical dependence of the electron energy of an electron state on a distance between nuclei in a diatomic molecule

Diatomic molecules

Legend (key):

- Ionization potential (eV)
- Dissociation energy, eV
- Symbol
- Atomic number
- Element
- Electron term
- Vibrational energy, cm^{-1}
- Anharmonic constant, cm^{-1}
- Rotational constant, cm^{-1}
- Equilibrium distance, Å

Example: 6.4 ; 2.62 ; $_{23}$V ; Vanadium ; $^3\Sigma_g^+$; 537.5 ; 1.78 ; 0.209

Group / Period	I	II	III	IV	V	VI	VII	VIII
1	15.43 ; 4.478 $_1$H $^1\Sigma_g^+$ 4401 0.741 121.3 Hydrogen 60.85 0.96	22.22 ; 0.001 $_2$He $^1\Sigma_g^+$ 2.97 Helium						
2	5.145 ; 1.05 $_3$Li $^1\Sigma_g^+$ 351.4 0.741 2.59 Lithium 0.672	7.45 ; 0.098 $_4$Be $^1\Sigma_g^+$ 2.45 275.8 13.5 Beryllium 0.615	8.8 : 2.8 $_5$B $^?\Sigma$ 1.60 Boron	12.15 ; 5.36 $_6$C $^1\Sigma_g^+$ 1.24 1855 13.17 Carbon 1.899	15.581 ; 9.579 $_7$N $^1\Sigma_g^+$ 2359 1.098 14.95 Nitrogen 1.998	12.071 ; 5.12 $_8$O $^3\Sigma_g^-$ 1580 1.207 11.98 Oxygen 1.445	15.686 ; 1.66 $_9$F $^1\Sigma_g^+$ 916.6 1.41 11.24 Fluorine 0.89	20.4 ; 0.037 $_{10}$Ne $^1\Sigma_g^+$ 3.09 Neon
3	4.90 ; 0.731 $_{11}$Na $^1\Sigma_g^+$ 159.1 3.08 1.62 Sodium 0.155	6.7 ; 0.053 $_{12}$Mg $^1\Sigma_g^+$ 3.89 51.08 0.725 Magnesium 0.093	4.84 : 0.46 $_{13}$Al $^3\Pi_u$ 2.47 284.2 1.02 Aluminium 0.205	7.4 ; 3.24 $_{14}$Si $^3\Sigma_g^-$ 2.24 510.9 1.02 Silicon 0.239	10.56 $_{15}$P $^1\Sigma_g^+$ 780.8 1.89 2.83 Phosphorus 0.304	9.4 ; 4.37 $_{16}$S $^3\Sigma_g^-$ 725.6 1.89 2.28 Sulfur 0.295	11.50 ; 2.576 $_{17}$Cl $^1\Sigma_g^+$ 559.7 1.99 1.68 Chlorine 0.244	14.54 ; 0.012 $_{18}$Ar $^1\Sigma_g^+$ 3.76 30.68 2.42 Argon 0.060
4	4.064 ; 0.551 $_{19}$K $^1\Sigma_g^+$ 92.09 3.92 0.283 Potassium 0.165	5.2 ; 0.13 $_{20}$Ca $^1\Sigma_g^+$ 64.9 4.28 1.09 Calcium 0.047	1.69 $^5\Sigma_u^-$ $_{21}$Sc 2.21 238.9 0.93 Scandium 0.153	6.2 ; 1.4 $_{22}$Ti $^3\Delta_g$ 407.9 1.94 1.03 Titanium 0.187	6.4 ; 2.62 $_{23}$V $^3\Sigma_g^+$ 537.5 1.78 0.209 Vanadium	6.8 ; 1.66 $_{24}$Cr $^1\Sigma_g^+$ 470 1.68 14.1 Chromium 0.23	6.5 ; 0.79 $_{25}$Mn $^1\Sigma_g^+$ 68.1 2.52 1.05 Manganese 0.097	6.3 ; 0.9 $_{26}$Fe $^?\Sigma_g$ 2.02 1.4 Iron 0.148
	7.89 ; 1.99 $_{29}$Cu $^1\Sigma_g^+$ 2.21 Copper 0.022	9.0 ; 0.034 $_{30}$Zn $^1\Sigma_g^+$ 4.8 Zinc	6.5 : 1.18 $_{31}$Ga $^3\Pi_u$ 2.76 Gallium 0.063	7.2 : 2.5 $_{32}$Ge $^3\Sigma_g^-$ 2.44 259 Germanium 0.073	9.81 ; 3.96 $_{33}$As $^1\Sigma_g^+$ 429.6 2.103 1.12 Arsenic 0.101	8.88 ; 2.9 $_{34}$Se $^3\Sigma_g^-$ 385.3 2.16 0.963 Selenium 0.89	10.52 ; 2.05 $_{35}$Br $^1\Sigma_g^+$ 325 2.28 1.08 Bromine 0.082	12.97 ; 0.017 $_{36}$Kr $^1\Sigma_g^+$ 24.1 4.01 1.34 Krypton 0.025
5	3.45 ; 0.495 $_{37}$Rb $^1\Sigma_g^+$ 246.4 4.17 Rubidium 0.109	4.74 ; 0.13 $_{38}$Sr $^1\Sigma_g^+$ 39.6 4.45 0.45 Strontium 0.019	1.6 $_{39}$Y $^1\Sigma_g^+$ 158 2.8 1.0 Yttrium	1.5 $_{40}$Zr $^3\Sigma_g^-$ 2.3 259 0.8 Zirconium 0.048	6.37 ; 5.48 $_{41}$Nb $^3\Sigma_g^-$ 424.9 2.1 373 Niobium 0.070	6.2 ; 4.1 $_{42}$Mo $^1\Sigma_g^+$ 477 2.2 0.94 Molybdenum 0.072	9.3 ; 1.542 $_{53}$I $^1\Sigma_g^+$ 214.5 2.67 0.615 Iodine 0.037	11.13 ; 0.024 $_{54}$Xe $^1\Sigma_g^+$ 21.13 4.36 0.65 Xenon 0.0135
	7.66 ; 1.65 $_{47}$Ag $^1\Sigma_g^+$ 192.4 2.53 0.4 Silver 0.011	0.013 $_{48}$Cd $^1\Sigma_g^+$ 5.1 Cadmium	0.83 $_{49}$In $^3\Pi_u$ 3.14 186.2 0.3 Indium 0.030	7.32 ; 2.0 $_{50}$Sn $^3\Sigma_g^-$ 2.75 Tin 0.261 0.038	8.8 ; 3.09 $_{51}$Sb $^1\Sigma_g^+$ 269.9 2.34 0.53 Antimony 0.050	8.29 ; 2.7 $_{52}$Te $^3\Sigma_g^-$ 249.1 2.56 0.537 Tellurium 0.040		
6	3.64 ; 0.452 $_{55}$Cs $^1\Sigma_g^+$ 42.02 4.65 0.013 Cesium 0.082	9.4 ; 0.055 $_{56}$Ba $^1\Sigma_g^+$ 42.1 4.6 0.16 Barium 0.069	6.3 ; 0.001 $_{81}$Tl $^3\Sigma_g^-$ 110.2 3.0 0.5 Thallium 0.013	6.4 : 0.83 $_{82}$Pb $^3\Sigma_g^-$ 110.2 2.93 0.327 Lead 0.019	7.4 : 2.08 $_{83}$Bi $^1\Sigma_g^+$ 173.1 2.66 0.376 Bismuth 0.023	6.9 $_{84}$Po Polonium		
	9.2 ; 2.31 $_{79}$Au $^1\Sigma_g^+$ 190.9 2.47 0.42 Gold 0.028	9.4 ; 0.055 $_{80}$Hg $^1\Sigma_g^+$ 18.5 3.65 0.27 Mercury 0.013	$_{57}$La Lanthanum	$_{58}$Hf Hafnium	$_{73}$Ta 7.4 Tantalum	7.4 $_{74}$W 316.8 Tungsten 1.0	$_{75}$Re Rhenium	$_{76}$Os Osmium

Right-side column (Periods 4–6, Group VIII transition metals):

6.0 : 0.9 $_{27}$Co $^1\Sigma_g^+$ 280 2.0 - Cobalt 0.14	7.43 : 1.7 $_{28}$Ni $^1\Sigma_g^+$ 350 2.3 1.1 Nickel 0.104
1.5 $_{45}$Rh $^5\Delta_g$ 2.67 238 Rhodium 0.046	7.7 ; 0.76 $_{46}$Pd $^3\Delta_g$ 2.48 159 Palladium 0.051
$_{44}$Ru Ruthenium	$_{43}$Tc Technetium
$_{77}$Ir Iridium	8.7 : 0.93 $_{78}$Pt $^1\Sigma_g^+$ 259.4 2.34 0.0 Platinum 0.032
	0.030 $_{86}$Rn $^1\Sigma_g^+$ 4.68 = Radon 0.0069
$_{85}$At Astatine	

Fig. 5.5 Parameters of homonuclear diatomic molecules

Positive diatomic

Term: $^2\Sigma_g^+$ (S)

Legend (cell key):

Diatomic ionization potential (eV)	4.90; 0.98
Symbol	Na
Atomic number	11
Element	Sodium
Electron term	$^2\Sigma_g^+$
Dissociation energy, eV	4.90
Vibrational energy, cm^{-1}	120.8
Anharmonic constant, cm^{-1}	0.46
Rotational constant, cm^{-1}	0.118
Equilibrium distance, Å	3.54

Parameters of homonuclear positive diatomic ions (as printed):

Z	Symbol	Element	Group	Period	Ion. pot. (eV)	Term	Other printed values
1	H	Hydrogen	I	1	15.43; 2.65	$^2\Sigma_g^+$	2313; 1.06; 67.5; 35.3
2	He	Helium	II	1	22.22; 2.47	$^2\Sigma_u^+$	1698.5; 1.08; 7.21
3	Li	Lithium	I	2	5.145; 1.28	$^2\Sigma_g^+$	263.1; 3.12; 1.60; 0.49
4	Be	Beryllium	II	2	7.45; 1.9	$^2\Sigma_g^+$	501; 4.2; 0.752
5	B	Boron	III	2	8.8; 1.9	$^2\Sigma_g^+$	357; 4.15; 0.181
6	C	Carbon	IV	2	12.15; 5.3		1.41
7	N	Nitrogen	V	2	15.581; 8.713	$^2\Sigma_g^+$	2207; 1.12; 16.2; 1.932
8	O	Oxygen	VI	2	12.071; 6.66	$^2\Pi_g$	1905; 1.12; 16.3; 1.689
9	F	Fluorine	VII	2	15.47; 3.34	$^2\Pi_g$	1073; 1.32; 9.13; 1.015
10	Ne	Neon	VIII	2	20.4; 1.2	$^2\Sigma_u^+$	586; 1.75; 5.4; 0.344
11	Na	Sodium	I	3	4.90; 0.98	$^2\Sigma_g^+$	120.8; 3.54; 0.46; 0.118
12	Mg	Magnesium	II	3	6.7; 1.3	$^2\Sigma_g^+$	
13	Al	Aluminium	III	3	4.84; 1.4	$^2\Sigma_g^+$	3.2; 2.0; 0.122
14	Si	Silicon	IV	3	7.4; 3.24	$^2\Pi_u$	528; 2.19; 0.25
15	P	Phosphorus	V	3	10.56; 5.0	$^2\Pi_u$	672; 1.98; 2.74; 0.276
16	S	Sulfur	VI	3	9.4; 5.4	$^2\Pi_u$	806; 1.82; 3.33; 0.318
17	Cl	Chlorine	VII	3	11.50; 3.95	$^2\Pi_g$	645.6; 1.88; 3.02; 0.265
18	Ar	Argon	VIII	3	14.54; 1.23	$^2\Sigma_u^+$	308.9; 2.43; 1.66; 0.143
19	K	Potassium	I	4	4.064; 0.81	$^2\Sigma_g^+$	73.4; 4.6; 0.2; 0.042
20	Ca	Calcium	II	4	5.2; 1.04	$^2\Sigma_g^+$	119; 3.7; 0.053
21	Sc	Scandium	III	4			
22	Ti	Titanium	IV	4	6.0; 2.44		
23	V	Vanadium	V	4	6.36; 3.14		
24	Cr	Chromium	VI	4	6.8; 1.8		
25	Mn	Manganese	VII	4	6.47; 1.3		
26	Fe	Iron	VIII	4	6.3; 2.7		
27	Co	Cobalt			6.0; 2.75		
28	Ni	Nickel			7.43; 2.35		
29	Cu	Copper	I	4	7.89; 1.8	$^2\Sigma_g^+$	183; 0.75; 0.096
30	Zn	Zinc	II	4	9.0; 0.42	$^2\Sigma_g^+$	0.54; 0.025
31	Ga	Gallium	III	4	6.5; 1.27	$^2\Sigma_g^+$	108; 3.24; 0.046
32	Ge	Germanium	IV	4	7.2; 2.91	$^2\Sigma_g^+$	156; 2.32; 0.026
33	As	Arsenic	V	4	9.64; 4.11	$^2\Sigma_g^+$	430; 1.9; 0.125
34	Se	Selenium	VI	4	8.88; 4.4	$^2\Pi_u$	450; 2.07; 0.10
35	Br	Bromine	VII	4	10.52; 2.96	$^2\Pi_g$	376; 2.3; 1.13; 0.033
36	Kr	Krypton	VIII	4	12.97; 1.15	$^2\Sigma_u^+$	178; 2.8; 0.82; 0.051
37	Rb	Rubidium	I	5	3.45; 0.75	$^2\Sigma_g^+$	44.5; 4.8; 0.017
38	Sr	Strontium	II	5	4.74; 1.1	$^2\Sigma_g^+$	3.9; 0.017
39	Y	Yttrium	III	5			
40	Zr	Zirconium	IV	5			
41	Nb	Niobium	V	5	6.37; 5.87		
42	Mo	Molybdenum	VI	5	6.2; 5.0		
43	Tc	Technetium	VII	5			
44	Ru	Ruthenium	VIII	5			
45	Rh	Rhodium					
46	Pd	Palladium			7.7		
47	Ag	Silver	I	5	7.66; 1.69	$^2\Sigma_g^+$	113; 2.8; 0.05; 0.040
48	Cd	Cadmium	II	5		$^2\Sigma_g^+$	
49	In	Indium	III	5			
50	Sn	Tin	IV	5	7.38; 1.96		
51	Sb	Antimony	V	5	8.7; 3.2		
52	Te	Tellurium	VI	5	8.2; 3.5		
53	I	Iodine	VII	5	9.3; 1.92	$^2\Pi_g$	143; 2.58; 0.040
54	Xe	Xenon	VIII	5	11.13; 1.1	$^2\Sigma_u^+$	123; 3.25; 0.63; 0.026
55	Cs	Cesium	I	6	3.76; 0.61	$^2\Sigma_g^+$	32.4; 4.44; 0.05; 0.013
56	Ba	Barium	II	6		$^2\Sigma_g^+$	
57	La	Lanthanum	III	6			
58	Hf	Hafnium	IV	6			
73	Ta	Tantalum	V	6			
74	W	Tungsten	VI	6			
75	Re	Rhenium	VII	6			
76	Os	Osmium	VIII	6			
77	Ir	Iridium					
78	Pt	Platinum			8.7; 3.26		
79	Au	Gold	I	6	9.2	$^2\Sigma_g^+$	91.6; 0.301; 0.021
80	Hg	Mercury	II	6	9.4; 0.96	$^2\Sigma_g^+$	2.8; 0.021
81	Tl	Thallium	III	6			
82	Pb	Lead	IV	6	6.1; 1.7	$^2\Pi_g$	143.4; 0.18
83	Bi	Bismuth	V	6	7.4; 1.89		
84	Po	Polonium	VI	6			
85	At	Astatine	VII	6			
86	Rn	Radon	VIII	6			

Fig. 5.6 Parameters of homonuclear positive diatomic ions

Negative diatomic homonuclear ions

Period / Group	I	II	III	IV	V	VI	VII
2	0.7; 0.88 / $^2\Sigma_u^+$ / 233.1 / 1.92 / 0.516 / $_{3}$Li 2.8 / Lithium	0.4; 2.8 / $^2\Sigma_g^+$ / 620 / — / $_{4}$Be 2.3 / Beryllium	1.7; 4.2 / $_{5}$B / Boron	3.27; 8.5 / $^2\Sigma_g^+$ / 1781 / 11.7 / 1.746 / $_{6}$C 1.27 / Carbon	$_{7}$N / Nitrogen	0.45; 4.16 / $^2\Pi_g$ / 1090 / 10 / 1.12 / $_{8}$O 1.35 / Oxygen	3.08; 1.3 / $^2\Sigma_u^+$ / 475 / 5.1 / 0.47 / $_{9}$F 1.92 / Fluorine
3	0.43; 0.44 / $^2\Sigma_u^+$ / 28.3 / 1.92 / — / $_{11}$Na / Sodium	$_{12}$Mg / Magnesium	1.1; 2.4 / $_{13}$Al 2.65 / Aluminium	2.19; 3.3 / $_{14}$Si / Silicon	0.59; 4.8 / $^2\Pi_u$ / 640 / 1.98 / 0.277 / $_{15}$P / Phosphorus	1.67; 3.95 / $^2\Pi_u$ / 601 / 2.16 / 0.32 / $_{16}$S 1.8 / Sulfur	2.38; 1.26 / $^2\Sigma_u^+$ / 277 / 1.8 / — / $_{17}$Cl / Chlorine
4 (even)	0.49; / $_{19}$K / Potassium	$_{20}$Ca / Calcium	$_{21}$Sc / Scandium	$_{22}$Ti / Titanium	0.1; 2.7 / $^2\Pi_g$ / 330 / 0.86 / $_{23}$V / Vanadium	; 0.17 / $_{24}$Cr / Chromium	$_{25}$Mn / Manganese
4 (odd)	0.84; 1.57 / $^2\Sigma_u^+$ / 196 / 0.7 / 0.097 / $_{29}$Cu 2.34 / Copper	$_{30}$Zn / Zinc	6.5; 1.18 / $^4\Sigma_g^-$ / 153 / 1.0 / 0.063 / $_{31}$Ga 2.76 / Gallium	$_{32}$Ge / Germanium	$_{33}$As / Arsenic	1.94; / $_{34}$Se / Selenium	2.55; 1.2 / $^2\Sigma_u^+$ / 175 / 0.33 / 0.054 / $_{35}$Br 2.81 / Bromine
5 (even)	0.5; 0.5 / $^2\Sigma_u^+$ / 28.3 / $_{37}$Rb 4.8 / Rubidium	$_{38}$Sr / Strontium	$_{39}$Y / Yttrium	$_{40}$Zr / Zirconium	$_{41}$Nb / Niobium	$_{42}$Mo / Molybdenum	$^2\Sigma_u^+$ / 175 / 0.33 / 0.054 / $_{43}$Tc / Technetium
5 (odd)	1.03; 1.37 / $^2\Sigma_u^+$ / — / 0.042 / 0.017 / $_{47}$Ag 2.6 / Silver	$_{48}$Cd / Cadmium	$_{49}$In / Indium	$_{50}$Sn / Tin	$_{51}$Sb / Antimony	1.92; / $_{52}$Te / Tellurium	2.55; 1.05 / $^2\Sigma_u^+$ / $_{53}$I / Iodine
6 (even)	0.47; 0.45 / $^2\Sigma_u^+$ / 28.4 / 0.9 / 0.046 / $_{55}$Cs 4.8 / Cesium	$_{56}$Ba / Barium	$_{57}$La / Lanthanum	$_{58}$Hf / Hafnium	1.27; 2.8 / $^2\Pi_g$ / 152 / 0.53 / 0.020 / $_{73}$Ta / Tantalum	6.9 / 336.8 / 1.0 / $_{74}$W / Tungsten	$_{75}$Re / Rhenium
6 (odd)	1.94; 1.9 / $^2\Sigma_u^+$ / $_{79}$Au / Gold	3.65 / $_{80}$Hg / Mercury	$_{81}$Tl / Thallium	1.66; 1.37 / $^2\Pi_u$ / 129 / 0.2 / 0.021 / $_{82}$Pb 2.81 / Lead	1.27; 2.8 / 2.83 $_{83}$Bi / Bismuth	$_{84}$Po / Polonium	$_{85}$At / Astatine

Legend:
- Dissociation energy, eV
- Electron affinity, eV — 0.47; 0.45
- Electron term of the ground state — $^2\Sigma_u^+$
- Symbol — Cs
- Atomic number — 55
- Element — Cesium
- Vibration energy, cm^{-1} — 4.8
- Rotation constant, cm^{-1} — 0.042
- Anharmonicity constant, cm^{-1} — 0.011
- Equilibrium distance, Å

Fig. 5.7 Parameters of homonuclear negative diatomic ions

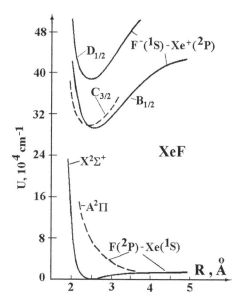

Fig. 5.8 Potential curves of excimer molecule XeF

Excimer molecules.

	Ne	Ar	Kr	Xe
F	$B_{1/2}$ $^{2.00\ 6.41}$ [108, 2.5] $C_{3/2}$ $^{1.99\ 6.35}$	$B_{1/2}$ [193, 4]	$B_{1/2}$ $^{2.51\ 5.30}$ [248, 8] $C_{3/2}$ $^{2.44\ 5.21}$ $D_{1/2}$ $^{2.47\ 5.26}$	$X_{1/2}$ $^{2.29\ 0.15}$ $B_{1/2}$ $^{2.63\ 5.30}$ [352, 16] $C_{3/2}$ $^{2.56\ 5.03}$ [450, 100] $D_{1/2}$ $^{2.51\ 5.46}$ [260, 11]
Cl		$B_{1/2}$ [175, 9]	$B_{1/2}$ [222, 19]	$X_{1/2}$ $^{3.2\ 0.032}$ $B_{1/2}$ $^{3.22\ 4.23}$ [308, 11] $C_{3/2}$ $^{3.14\ 4.14}$ [330, 120] $D_{1/2}$ $^{3.18\ 4.17}$ [236, 10]
Br		$B_{1/2}$ $^{2.81\ 4.74}$		$B_{1/2}$ $^{3.38\ 4.30}$ [282, 15] $C_{3/2}$ $^{3.31\ 3.95}$ [302, 120] $D_{1/2}$ $^{3.34\ 3.98}$ [221, 9]
I		Distance between nuclei at minimum, Å — Minimum of the potential energy, eV Electron state — $B_{1/2}$ $^{3.38\ 4.30}$ [282, 15] $C_{3/2}$ $^{3.31\ 3.95}$ [302, 120] $D_{1/2}$ $^{3.34\ 3.98}$ [221, 9] — Wavelength for the band center of the radiative transition, nm and radiative lifetime, ns		$B_{1/2}$ $^{3.62\ 4.08}$ [254, 14] $C_{3/2}$ $^{3.57\ 3.71}$ [292, 110] $D_{1/2}$ $^{3.59\ 3.75}$ [203, 9]

Fig. 5.9 Parameters of excimer molecules

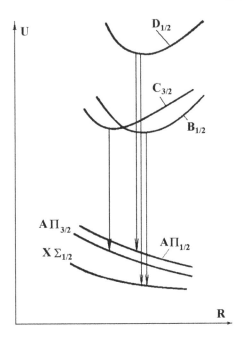

Fig. 5.10 Character of radiative transitions in an excimer molecule

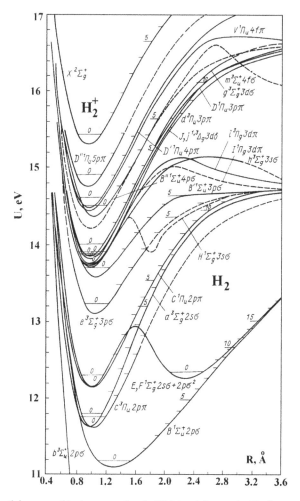

Fig. 5.11 Potential curves of hydrogen molecule (H₂) involving excited hydrogen atoms

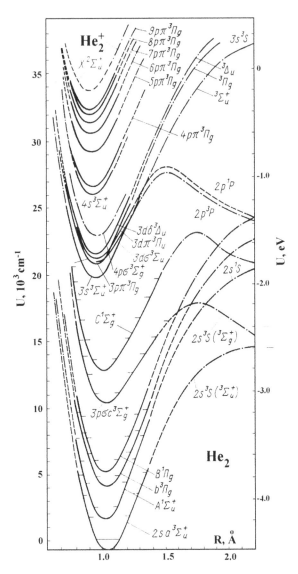

Fig. 5.12 Potential curves of helium molecule (He$_2$) involving excited helium atoms. The energy of electron terms expressed in cm^{-1} starts from the ground vibration and lowest electron excited state, the energy of electron terms in eV begins from the ground vibration and electron states

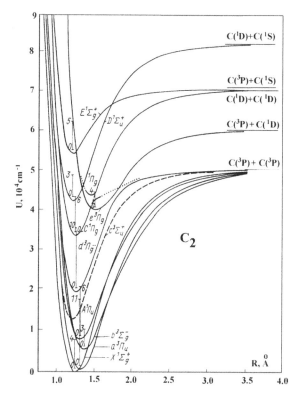

Fig. 5.13 Potential curves of the carbon diatomic molecule C_2

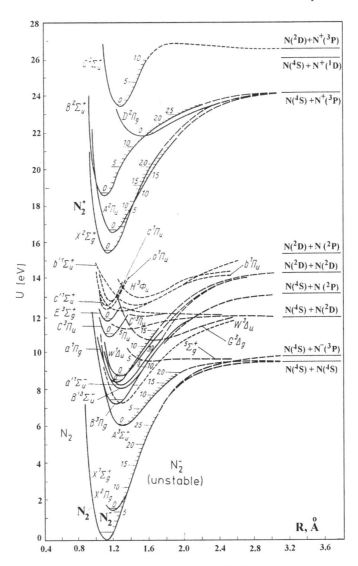

Fig. 5.14 Potential curves of the nitrogen diatomic molecule N$_2$

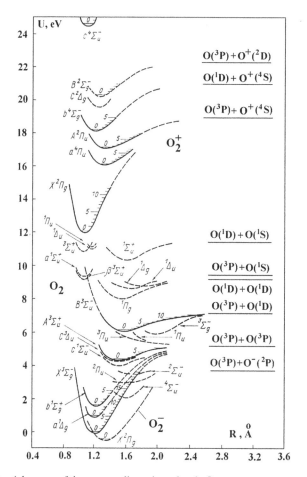

Fig. 5.15 Potential curves of the oxygen diatomic molecule O_2

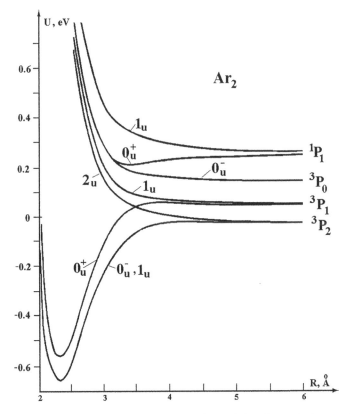

Fig. 5.16 Potential curves of the argon diatomic molecule consisting of atoms in the ground Ar($2p^6$) and lowest excited Ar($2p^5 3s$) states which start from the lowest excited atom state at large distances from the argon atom in the ground state

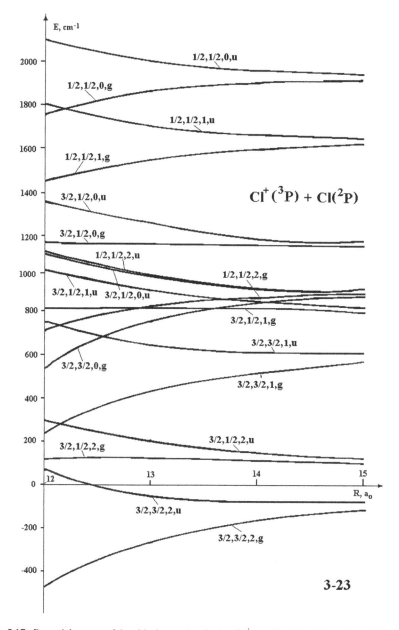

Fig. 5.17 Potential curves of the chlorine molecular ion Cl_2^+ resulted form interaction of the atom $Cl(^2P)$ and ion $Cl^+(^3P)$ in the lowest electron states at large separations which are responsible for the resonant charge exchange process [100]. The excitation energies are counted off form the ground ion and atom states at infinite distance between them and are expressed in cm^{-1}. The indicated quantum number of electron terms are $JM_J j$ and the evenness, where J, j are the total momenta of the atom and ion, M_J is the projection of the total atom momentum onto the molecular axis

Electron affinity of molecules.

Legend (example cell):

- $EA(A_3)$, eV
- $EA(A_2)$, eV
- $EA(A_0)$, eV — Symbol, Atomic weight
- $EA(A_1)$, eV — Element

$EA(A_3)$	$EA(A_2)$	Symbol (Atomic weight)
2.09	1.67	$_{16}$S (Sulfur)
1.125	2.314	Element

Main table

Group Period	I	II	III	IV	V	VI	VII
2	0.6 $_3$Li 0.34 Lithium	$_4$Be 0.7 Beryllium	2.51 $_5$B Boron	1.98 3.27 $_6$C Carbon	2.7 0.03 0.35 $_7$N Nitrogen	2.10 0.45 $_8$O 1.828 Oxygen	2.27 3.1 $_9$F Fluorine
3	0.43 $_{11}$Na 1.02 Sodium	$_{12}$Mg 1.63 1.05 Magnesium	1.1 1.4 $_{13}$Al 2.60 Aluminium	2.29 2.20 $_{14}$Si 1.28 Silicon	0.59 $_{15}$P 1.09 Phosphorus	2.09 1.67 $_{16}$S 1.125 2.314 Sulfur	2.28 2.4 $_{17}$Cl Chlorine
4	0.50 $_{19}$K 0.96 Potassium	$_{20}$Ca 0.96 0.93 Calcium	$_{21}$Sc 1.35 Scandium	2.23 $_{22}$Ti 1.30 Titanium	1.11 1.23 $_{23}$V Vanadium	0.505 $_{24}$Cr 1.22 0.56 Chromium	$_{25}$Mn 1.38 0.87 Manganese
4 (sub)	0.84 $_{29}$Cu 1.78 Copper	2.09 $_{30}$Zn 0.95 Zinc	$_{31}$Ga 1.35 Gallium	2.035 $_{32}$Ge Germanium	1.45 0.74 $_{33}$As 1.29 1.0 Arsenic	1.46 1.9 $_{34}$Se 2.212 Selenium	2.35 $_{35}$Br 2.55 Bromine
5	0.50 $_{37}$Rb 2.3 Rubidium	$_{38}$Sr 0.92 Strontium	$_{39}$Y 1.35 Yttrium	$_{40}$Zr 0.60 1.3 Zirconium	1.03 1.29 $_{41}$Nb 1.85 Niobium	$_{42}$Mo 1.29 Molybdenum	4.23 $_{43}$Tc 2.55 Technetium
5 (sub)	1.02 $_{47}$Ag Silver	$_{48}$Cd 1.02 0.27 Cadmium	$_{49}$In Indium	1.96 $_{50}$Sn Tin	$_{51}$Sb Antimony	1.70 1.9 $_{52}$Te 2.10 Tellurium	2.38 $_{53}$I 2.55 Iodine
6	0.47 $_{55}$Cs 3.7 Cesium	$_{56}$Ba 0.86 Barium	$_{57}$La 1.4 Lanthanum	$_{72}$Hf 0.72 0.60 Hafnium	$_{73}$Ta 1.36 1.60 Tantalum	$_{74}$W Tungsten	1.57 $_{75}$Re Rhenium
6 (sub)	1.94 $_{79}$Au Gold	$_{80}$Hg Mercury	$_{81}$Tl Thallium	1.37 $_{82}$Pb Lead	1.27 $_{83}$Bi Bismuth		

Transition metal triads (shown separately)

0.90 1.5 $_{26}$Fe 1.49 0.93 Iron	1.11 $_{27}$Co 0.67 Cobalt	0.93 1.4 $_{28}$Ni 1.470 0.48 Nickel
0.90 $_{44}$Ru 1.29 Ruthenium	1.11 1.58 $_{45}$Rh Rhodium	1.68 1.5 $_{46}$Pd Palladium
0.90 $_{76}$Os Osmium	1.90 $_{77}$Ir Iridium	1.90 1.87 $_{78}$Pt Platinum

Fig. 5.18 Electron affinity of molecules

Fig. 5.19 Potential curves
of the CH molecule

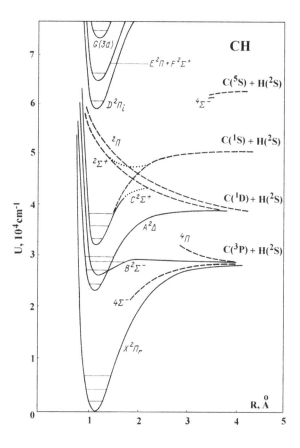

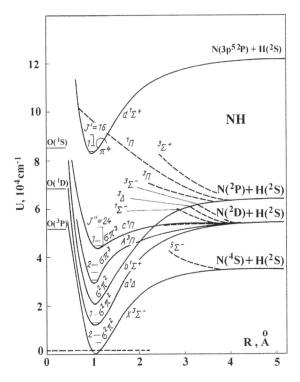

Fig. 5.20 Potential curves of molecule NH

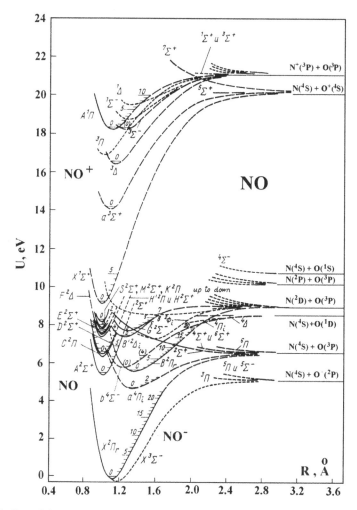

Fig. 5.21 Potential curves of molecule OH

Atom affinity to hydrogen and oxygen atoms

Legend:

- Shell of valence electrons
- Symbol
- Atomic number
- Element
- Electron term of the ground state in L-S scheme notations
- Affinity to hydrogen atom in eV
- Affinity to oxygen atom in eV

Example: 4s² — Ca — 20 — Calcium — ¹S₀ — [1.74] [4.2]

The affinity of atom M to the hydrogen or oxygen atom is the dissociation energy of radical MH or MO in the ground vibration state.

Group	Config	Element	Z	Name	Term	Affinity to H (eV)	Affinity to O (eV)
I	1s	H	1	Hydrogen	$^2S_{1/2}$	4.48	4.39
I	2s	Li	3	Lithium	$^2S_{1/2}$	2.47	3.5
I	3s	Na	11	Sodium	$^2S_{1/2}$	1.92	2.6
I	4s	K	19	Potassium	$^2S_{1/2}$	1.81	2.9
I	3d¹⁰4s	Cu	29	Copper	$^2S_{1/2}$	2.9	2.8
I	5s	Rb	37	Rubidium	$^2S_{1/2}$	1.7	2.8
I	4d¹⁰5s	Ag	47	Silver	$^2S_{1/2}$	2.2	2.3
I	6s	Cs	55	Cesium	$^2S_{1/2}$	1.78	3.0
I	5d¹⁰6s	Au	79	Gold	$^2S_{1/2}$	3.1	2.3
II	2s²	Be	4	Beryllium	$^1S_0$	2.05	4.5
II	3s²	Mg	12	Magnesium	$^1S_0$	1.31	3.8
II	4s²	Ca	20	Calcium	$^1S_0$	1.74	4.2
II	4s²	Zn	30	Zinc	$^1S_0$	0.89	4.7
II	5s²	Sr	38	Strontium	$^1S_0$	1.7	1.6
II	5s²	Cd	48	Cadmium	$^1S_0$	1.7	4.4
II	6s²	Ba	56	Barium	$^1S_0$	0.7	2.4
II	5d¹⁰6s²	Hg	80	Mercury	$^1S_0$	0.41	2.3
III	2p	B	5	Boron	^2P	3.5	8.4
III	3p	Al	13	Aluminium	$^2P_{1/2}$	2.95	5.25
III	3d4s²	Sc	21	Scandium	$^2D_{3/2}$	3.5	7.8
III	4p	Ga	31	Gallium	$^2P_{1/2}$	3.3	7.5
III	4d5s²	Y	39	Yttrium	$^2D_{3/2}$	2.8	3.8
III	5p	In	49	Indium	$^2P_{1/2}$	2.7	5.5
III	5d6s²	La	57	Lanthanum	$^2D_{3/2}$	1.9	3.3
III	6p	Tl	81	Thallium	$^2P_{1/2}$	1.9	4.0
IV	2p²	C	6	Carbon	$^3P_0$	3.46	11.09
IV	3p²	Si	14	Silicon	$^3P_0$	3.1	8.3
IV	3d²4s²	Ti	22	Titanium	$^3F_2$	2.1	7.0
IV	4p²	Ge	32	Germanium	$^3P_0$	3.3	6.8
IV	4d²5s²	Zr	40	Zirconium	$^3F_2$	2.7	5.5
IV	5p²	Sn	50	Tin	$^3P_0$	2.7	5.5
IV	5d²6s²	Hf	72	Hafnium	$^3F_2$	1.6	3.5
IV	6p²	Pb	82	Lead	$^3P_0$	3.0	8.3
V	2p³	N	7	Nitrogen	$^4S_{3/2}$	3.5	6.50
V	3p³	P	15	Phosphorus	$^4S_{3/2}$	3.1	6.1
V	3d³4s²	V	23	Vanadium	$^4F_{3/2}$	2.2	6.5
V	4p³	As	33	Arsenic	$^4S_{3/2}$	2.84	4.95
V	4d⁴5s	Nb	41	Niobium	$^6D_{1/2}$		8.0
V	5p³	Sb	51	Antimony	$^4S_{3/2}$	4.5	
V	5d³6s²	Ta	73	Tantalum	$^4F_{3/2}$	2.7	
V	6p³	Bi	83	Bismuth	$^4S_{3/2}$		8.3
VI	2p⁴	O	8	Oxygen	$^3P_2$	4.39	5.12
VI	3p⁴	S	16	Sulfur	$^3P_2$	3.6	5.4
VI	3d⁵4s	Cr	24	Chromium	$^7S_3$	2.0	4.2
VI	4p⁴	Se	34	Selenium	$^3P_2$	3.2	4.8
VI	4d⁵5s	Mo	42	Molybdenum	$^7S_3$	2.78	3.9
VI	5p⁴	Te	52	Tellurium	$^3P_2$	2.78	6.0
VI	5d⁴6s²	W	74	Tungsten	$^5D_0$		7.0
VII	2p⁵	F	9	Fluorine	$^2P_{3/2}$	5.91	2.4
VII	3p⁵	Cl	17	Chlorine	$^2P_{3/2}$	4.47	2.75
VII	3d⁵4s²	Mn	25	Manganese	$^6S_{5/2}$	2.4	4.0
VII	4p⁵	Br	35	Bromine	$^2P_{3/2}$	3.79	2.44
VII	4d⁵5s²	Tc	43	Technetium	$^6S_{5/2}$	3.07	2.88
VII	5p⁵	I	53	Iodine	$^2P_{3/2}$	3.07	2.88
VII	5d⁵6s²	Re	75	Rhenium	$^6S_{5/2}$		6.5
VIII	1s²	He	2	Helium	$^1S_0$		
VIII	2p⁶	Ne	10	Neon	$^1S_0$	1.9	4.0
VIII	3p⁶	Ar	18	Argon	$^1S_0$		
VIII	3d⁶4s²	Fe	26	Iron	$^5D_4$	2.6	4.0
VIII	4p⁶	Kr	36	Krypton	$^1S_0$		
VIII	4d⁷5s	Ru	44	Ruthenium	$^5F_5$	2.4	5.5
VIII	5p⁶	Xe	54	Xenon	$^1S_0$		
VIII	5d⁶6s²	Os	76	Osmium	$^5D_4$		6.4
VIII	3d⁷4s²	Co	27	Cobalt	$^4F_{9/2}$	2.3	4.0
VIII	3d⁸4s²	Ni	28	Nickel	$^3F_4$	2.6	4.0
VIII	4d⁸5s	Rh	45	Rhodium	$^4F_{9/2}$	2.6	4.2
VIII	4d¹⁰	Pd	46	Palladium	$^1S_0$	2.4	4.0
VIII	5d⁷6s²	Ir	77	Iridium	$^4F_{9/2}$		4.3
VIII	5d⁹6s	Pt	78	Platinum	$^3D_3$	3.4	4.0

Fig. 5.22 Affinity of atoms to hydrogen and oxygen atoms

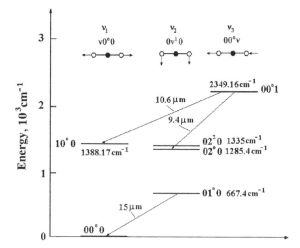

Fig. 5.23 Lowest vibration states of the CO_2 molecule and radiative transitions between them

Tables

Table 5.1 The quadrupole moment Q of atoms with filling p-shell for the LS-coupling scheme; M is the atom momentum projection onto the molecular axis

Atom states	$(p)^2 P$	$(p^2)^2 P$	$(p^3)^4 S$	$(p^4)^3 P$	$(p^5)^2 P$
$M = 0$	4/5	−4/5	0	4/5	−4/5
$|M| = 1$	−2/5	2/5	−	−2/5	2/5

Table 5.2 Given in atomic units the values of the constant C_6 of van der Waals interaction between two identical atoms of inert gases [87–89]

Interacting atoms	He-He	Ne-Ne	Ar-Ar	Kr-Kr	Xe-Xe
C_6	1.5	6.6	68	130	270

Table 5.3 The cases of Hund coupling [58, 59]

Hund case	Relation	Quantum numbers
a	$V_e \gg V_m \gg V_r$	Λ, S, S_n
b	$V_e \gg V_r \gg V_m$	Λ, S, S_N
c	$V_m \gg V_e \gg V_r$	Ω
d	$V_r \gg V_e \gg V_m$	L, S, L_N, S_N
e	$V_r \gg V_m \gg V_e$	J, J_N

Table 5.4 Parameters of interaction of a positive halogen ion with a parent atom in the ground electron states at a distance R_o that is responsible for resonant charge exchange at the collision energy of 1 eV [100]

	F	Cl	Br	I
R_o, a_o	10.6	13.8	15.1	17.2
δ_a, cm^{-1}	404	882	3685	7603
δ_i, cm^{-1}	490	996	3840	7087
V_{ex}, cm^{-1}	20873	11654	11410	13727
δ_i/V_{ex}	0.023	0.085	0.34	0.52
U_M, cm^{-1}	341	407	448	372
U_M/δ_a	0.84	0.46	0.12	0.049
V_{rot}, cm^{-1}	30	17	10	7.1
$\Delta(R_o)$, cm^{-1}	23	14	8.4	6.1

Table 5.5 Correlation between states of two identical atoms in the same electron states and the molecule consisting of these atoms in the case "a" of Hund coupling [108]. A number of molecule states of a given symmetry is indicated in parentheses if it is not one

Atomic states	Dimer states
^1S	$^1\Sigma_g^+$
^2S	$^1\Sigma_g^+, {}^3\Sigma_u^+$
^3S	$^1\Sigma_g^+, {}^3\Sigma_u^+, {}^5\Sigma_g^+$
^4S	$^1\Sigma_g^+, {}^3\Sigma_u^+, {}^5\Sigma_g^+, {}^7\Sigma_u^+$
^1P	$^1\Sigma_g^+(2), {}^1\Sigma_u^-, {}^1\Pi_g, {}^1\Pi_u, {}^1\Delta_g$
^2P	$^1\Sigma_g^+(2), {}^1\Sigma_u^-, {}^1\Pi_g, {}^1\Pi_u, {}^1\Delta_g, {}^3\Sigma_u^+(2), {}^3\Sigma_g^-, {}^3\Pi_g, {}^3\Pi_u, {}^3\Delta_u$
^3P	$^1\Sigma_g^+(2), {}^1\Sigma_u^-, {}^1\Pi_g, {}^1\Pi_u, {}^1\Delta_g, {}^3\Sigma_u^+(2), {}^3\Sigma_g^-, {}^3\Pi_g, {}^3\Pi_u, {}^3\Delta_u,$ $^5\Sigma_g^+(2), {}^5\Sigma_u^-, {}^5\Pi_g, {}^5\Pi_u, {}^5\Delta_g$
^4P	$^1\Sigma_g^+(2), {}^1\Sigma_u^-, {}^1\Pi_g, {}^1\Pi_u, {}^1\Delta_g, {}^3\Sigma_u^+(2), {}^3\Sigma_g^-, {}^3\Pi_g, {}^3\Pi_u, {}^3\Delta_u$ $^5\Sigma_g^+(2), {}^5\Sigma_u^-, {}^5\Pi_g, {}^5\Pi_u, {}^5\Delta_g, {}^7\Sigma_u^+(2), {}^7\Sigma_g^-, {}^7\Pi_g, {}^7\Pi_u, {}^7\Delta_g$
^1D	$^1\Sigma_g^+(3), {}^1\Sigma_u^-(2), {}^1\Pi_g(2), {}^1\Pi_u(2), {}^1\Delta_g(2), {}^1\Delta_u, {}^1\Phi_g, {}^1\Phi_u, {}^1\Gamma_g$
^2D	$^1\Sigma_g^+(3), {}^1\Sigma_u^-(2), {}^1\Pi_g(2), {}^1\Pi_u(2), {}^1\Delta_g(2), {}^1\Delta_u, {}^1\Phi_g, {}^1\Phi_u, {}^1\Gamma_g,$ $^3\Sigma_u^+(3), {}^3\Sigma_g^-(2), {}^3\Pi_g(2), {}^3\Pi_u(2), {}^3\Delta_g, {}^3\Delta_u(2), {}^3\Phi_g, {}^3\Phi_u, {}^3\Gamma_u$
^3D	$^1\Sigma_g^+(3), {}^1\Sigma_u^-(2), {}^1\Pi_g(2), {}^1\Pi_u(2), {}^1\Delta_g(2), {}^1\Delta_u, {}^1\Phi_g, {}^1\Phi_u, {}^1\Gamma_g,$ $^3\Sigma_u^+(3), {}^3\Sigma_g^-(2), {}^3\Pi_g(2), {}^3\Pi_u(2), {}^3\Delta_g, {}^3\Delta_u(2), {}^3\Phi_g, {}^3\Phi_u, {}^3\Gamma_u,$ $^5\Sigma_g^+(3), {}^5\Sigma_u^-(2), {}^5\Pi_g(2), {}^5\Pi_u(2), {}^5\Delta_g(2), {}^5\Delta_u, {}^5\Phi_g, {}^5\Phi_u, {}^5\Gamma_g$

Table 5.6 Correlation between states of two identical atoms in the same electron states and the molecule consisting of these atoms in the case "c" of Hund coupling [108]. A number of molecule states of a given symmetry is indicated in parentheses if it is not one

Atom states	Dimer states
$J = 0$	0_g^+
$J = 1/2$	$1_u, 0_g^+, 0_u^-$
$J = 1$	$2_g, 1_u, 1_g, 0_g^+(2), 0_u^-$
$J = 3/2$	$3_u, 2_g, 2_u, 1_g, 1_u(2), 0_g^+(2), 0_u^-(2)$
$J = 2$	$4_g, 3_g, 3_u, 2_g(2), 2_u, 1_g(2), 1_u(2), 0_g^+(3), 0_u^-(2)$

Table 5.7 Parameters of lowest excited states of inert gas molecules

Molecule, state	R_e, Å	D_e, eV	$\tau, 10^{-7}$, s
$He_2(a^3\Sigma_u^+)$	1.05	2.0	800
$He_2(A^1\Sigma_u^+)$	1.06	2.5	0.28
$Ne_2(a^3\Sigma_u^+)$	1.79	0.47	360
$Ar_2(1_u, 0_u^-)$	2.4	0.72	0.5
$Ar_2(0_u^+)$	2.4	0.69	30
$Kr_2(0_u^+)$			0.6
$Xe_2(1_u, 0_u^-)$	3.03	0.79	11
$Xe_2(0_u^+)$	3.02	0.77	0.6

Table 5.8 The distance R_h between nuclei and the hump height $\Delta\varepsilon$ for the interaction potential of excited and non-excited atoms of helium and neon

Molecule, state	R_h, Å	$\Delta\varepsilon$, eV
$He_2(a^3\Sigma_u^+)$	3.1	0.06
$He_2(A^1\Sigma_u^+)$	2.8	0.05
$Ne_2(a^3\Sigma_u^+)$	2.6	0.11
$Ne_2(A^1\Sigma_u^+)$	2.5	0.20

Table 5.9 The polarizabilities α of diatomic homonuclear molecules expressed in a_o^3

Molecule	α	Molecule	α
H_2	5.42	Al_2	130
Li_2	230	Cl_2	31
N_2	11.8	K_2	500
O_2	10.8	Br_2	474
F_2	93	Rb_2	530
Na_2	260	Cs_2	700

Table 5.10 Vibrational, rotational and structural parameters of three-atom molecule

Molecule	nu_1, cm^{-1}	nu_2, cm^{-1}	nu_3, cm^{-1}	A_o, cm^{-1}	B_o, cm^{-1}	C_o, cm^{-1}	r, Å	$\alpha,^o$
H_2O	3657	1595	3756	27.9	14.5	9.3	0.96	105
D_2O	2678	1178	2758	15.4	7.3	4.8	0.96	105
H_2S	2615	1183	2626	10.4	9.02	4.73	1.34	92
BO_2	1056	447	1278	–	0.328	–	1.26	180
CO_2	1333	667	2349	–	0.390	–	1.16	180
CF_2	1225	667	1114	2.95	0.413	0.365	1.30	105
CS_2	658	397	1535	–	0.109	–	1.55	180
F_2O	928	461	831	1.96	0.363	0.306	1.41	103
NH_2	3219	1497	3301	23.7	12.9	8.16	1.02	103
NO_2	1318	750	1618	8.00	0.434	0.410	1.19	134
O_3	1103	701	1042	3.55	0.445	0.395	1.28	117
SO_2	1151	518	1362	2.03	0.344	0.294	1.43	120
SF_2	855	345	870	0.838	0.307	0.228	1.59	98.3
SiH_2	2032	990	2022	8.10	7.02	3.70	1.52	92.1
SiF_2	855	345	870	1.021	0.294	0.228	1.59	101
XeF_2	515	213	555	–	0.114	–	1.98	180

Table 5.11 Binding energies for three-atom molecules XY_2

Molecule	J, eV	EA, eV	$D(XY - Y, \text{eV})$
H_2O	12.61	–	5.12
H_2S	10.46	–	3.9
BO_2	13.5	4.3	
CO_2	13.77		5.45
CF_2	10.42	0.16	
CS_2	10.07	0.90	4.51
F_2O	13.11		2.85
NH_2	11.14	0.77	3.9
NO_2	9.76	2.27	3.11
O_3	12.5	1	1.04
SO_2	12.3	1.11	5.66
SF_2	–	–	
SiH_2	8.9	1.12	
SiF_2	10.9	–	
XeF_2	12.4	–	

Table 5.12 Electron affinities of molecules AF_4 and AF_6 given in eV

A	AF_4	AF_6	A	AF_4	AF_6	A	AF_4	AF_6
S	1.9	0.8	Mo	–	4.0	Re	–	4.2
V	3.5	–	Ru			Os	3.9	5.0
Mn	5.4	–	Rh	5.4	4.8	Ir	4.3	5.4
Fe	5.4	–	Te	–	3.3	Pt	5.2	7.0
Se	–	3.0	W	–	3.7	U	1.7	5.0

Chapter 6
Elementary Processes in Gases and Plasmas

Abstract Elementary processes involving atoms, electrons and ions are analyzed. Peculiarities of cross sections and rate constants as characteristics of these processes are considered. Various pair and three body collision processes are considered as elastic electron-atom scattering, processes of excitation, quenching and ionization of atoms by electron impact, elastic scattering and resonant charge exchange for ion-atom collisions, atom ionization in collisions involving excited atoms, electron-ion recombination and electron attachment to molecules. Physics of these processes is analyzed and numerical parameters of their cross sections or rate constants are given. Additionally processes are represented in an excited or ionized air. The character of elastic electron-atom collisions including the Ramsauer effect is represented within the framework of the phase theory of scattering.

6.1 Parameters of Elementary Processes in Gases and Plasmas

Properties of rare atomic systems, gases and plasmas, are determined by various processes of collision of atomic particles [109]. In a general consideration of atomic collisions [110, 111], we use the cross section as a characteristic of collision of atomic particles. According to definition, the differential cross section of collision of two atomic particles $d\sigma$ that is introduced in the center-of-mass frame of reference, is a number of scattered particles per unit time per unit solid angle $d\Omega$ to the flux of incident particles. The parameters of elastic scattering of particles when the particle internal state is not changed, are given in Fig. 6.1 in the classical limit. The classical parameters of particle collision are the impact collision parameter ρ, a distance of closest approach r_o, and the scattering angle ϑ. If the interaction potential $U(R)$ of colliding particles is spherically symmetric (R is a distance between particles), the conservation of the angular momentum of particles in the course of collision leads to the following relation between the impact collision parameter and the distance of closest approach [112]

$$\frac{\rho^2}{r_o^2} = \left[1 - \frac{U(r_o)}{\varepsilon}\right], \tag{6.1}$$

© Springer International Publishing AG, part of Springer Nature 2018
B. M. Smirnov, *Atomic Particles and Atom Systems*, Springer Series on Atomic, Optical, and Plasma Physics 51, https://doi.org/10.1007/978-3-319-75405-5_6

where $\varepsilon = \mu v^2$ is the kinetic energy of colliding particles in the center-of-mass frame of reference, and μ is the reduced mass of colliding particles, v is their relative velocity. In the case of a monotonic dependence $\vartheta(\rho)$ of the scattering angle on the impact collision parameter the differential cross section of scattering in the classical limit is

$$d\sigma = 2\pi\rho\,d\rho \tag{6.2}$$

The model of hard spheres is a convenient model for scattering of classical particles in a strongly varied potential, and the interaction potential for this model corresponds to a solid wall of a radius R_o, as it is shown in Fig. 6.2. Within the framework of this model, when the cross section is independent of the collision velocity, the differential cross section $d\sigma$ and the transport (or diffusion) cross section $\sigma^* = \int(1 - \cos\vartheta)d\sigma$ are equal correspondingly (see Fig. 6.3)

$$d\sigma = \pi R_o^2\,d\cos\vartheta, \ \ \sigma^* = \pi R_o^2 \tag{6.3}$$

The transport cross section of scattering σ^* relates to transport of particles in a gas, whereas the total cross section $\sigma_t = \int d\sigma$ is the integral over the phase shift of colliding particles as a result of their collision. Because a shift of the phase of an electromagnetic wave which results from interaction of a radiating atomic particle with surrounding atomic particles has the same nature, the total cross section of scattering in collision of a radiating particle with gas particles may be responsible for broadening of spectral lines. The total cross section is infinite ($\hbar \to 0$) in the classical limit since weak scattering takes place at any large impact parameters of particle collisions.

The classical description holds true in the case where the main contribution to the scattering cross section is given by large collision momenta $l = \mu\rho v/\hbar \gg 1$. In the case of a spherically symmetric interaction potential between colliding particles, scattering for different collision momenta l proceeds independently, and any total parameters of scattering are the sum of these parameters for each collision momentum l. The characteristic of particle scattering is the scattering phase δ_l, that is the parameter of the wave function of colliding atomic particles at large distances between them. The connection between the differential cross section of scattering $d\sigma = 2\pi|f(\vartheta)|^2 d\cos\vartheta$, the scattering amplitude $f(\vartheta)$, the diffusion or transport cross section σ^* and the total scattering cross section σ_t have the form [46, 113, 114]

$$f(\vartheta) = \frac{1}{2iq}\sum_{l=0}^{\infty}(2l+1)(e^{2i\delta_l}-1)P_l(\cos\vartheta), \ \sigma^* = \frac{4\pi}{q^2}\sum_{l=0}^{\infty}(l+1)\sin^2(\delta_l-\delta_{l+1}), \ \sigma_t = \frac{4\pi}{q^2}\sum_{l=0}^{\infty}(2l+1)\sin^2\delta_l, \tag{6.4}$$

where q is the phase vector of colliding particles in the center-of-mass frame of reference, so that the energy of particles in a center-of-mass system is $\varepsilon = \hbar^2 q^2/2m_e$.

6.2 Collision Processes Involving Ions

Ions as a charged component of a plasma are of importance for plasma properties which result from collisions in a plasma involving ions [115, 117–119]. Ion-atom collisions with variation the ion momentum have a classical character, i.e. nuclei are moving in these collisions along classical trajectories. These processes include elastic ion-atom collisions and the charge exchange process. Elastic scattering of ions on atoms at low energies is determined by their interaction at large distances with the polarization interaction potential (5.5) between them. The peculiarity of the interaction potential (5.5) is the possibility of particle capture that means approach of colliding particles up to $R = 0$. The cross section of capture σ_{cap} for the polarization interaction potential is [112]

$$\sigma_{cap} = 2\pi \sqrt{\frac{\alpha e^2}{\mu g^2}}, \tag{6.5}$$

where μ is the reduced mass of colliding ion and atom, g is their relative velocity.

   In reality, the interaction potential differs from the polarization one (5.5) at distances compared with atomic sizes and includes ion-atom repulsion at small separations. Therefore the capture cross section (6.5) means only a strong approach of colliding ion and atom. Nevertheless, the diffusion cross section of ion-atom collision for the polarization interaction potential (5.5) between them σ_{ia}^* is close to the capture cross section (6.5) and is equal according to [116] $\sigma_{ia}^* = 1.10\sigma_{cap}$.

   Resonant charge exchange in ion collision with a parent atom proceeds according to the scheme

$$A^+ + \widetilde{A} \rightarrow \widetilde{A^+} + A, \tag{6.6}$$

where tilde marks one of colliding particles. Usually at room temperature and higher collision energies the cross section of resonant charge exchange exceeds remarkable the cross section of ion-atom elastic collision, and therefore transport of ions in the parent gas is determined by resonant charge exchange.

   Considering the resonant charge exchange process as result of state interference, we have for the electron wave function $\Psi(t)$ instead of formula (5.8)

$$\Psi(t) = \frac{\psi_g}{\sqrt{2}} \exp\left(-\frac{i}{2\hbar} \int_{-\infty}^{t} \varepsilon_g dt'\right) + \frac{\psi_u}{\sqrt{2}} \exp\left(-\frac{i}{2\hbar} \int_{-\infty}^{t} \varepsilon_u dt'\right), \tag{6.7}$$

where ε_g, ε_u are the energies of the even and odd states at a given distance between nuclei. From this one can find the probability of resonant charge exchange as a result of ion-atom collision [120]

$$P = |\langle \Psi(\infty)|\psi(1)\rangle|^2 = \sin^2 \zeta, \ \zeta = \frac{i}{2\hbar} \int_{-\infty}^{\infty} \Delta dt, \ \Delta = | \varepsilon_g - \varepsilon_u |, \qquad (6.8)$$

and ζ is the phase shift between even and odd states as a result of collision. This exhibits the interference nature of charge exchange, so that the electron state is the mixture of the even and odd states, and a phase shift between these states leads to the electron transition from the field of one core to another one. In turn, the phase shift is determined by the exchange interaction potential (5.8) in the course of collision.

The above expression (6.8) for the probability of the resonant charge process respects to a two-state approximation when the atom and ion electron states are not degenerated as it takes place if an atom A of the process (6.6) relates to elements of the first and second groups of the periodical table, i.e. for atoms with valence s-electrons. Nevertheless, this expression may be used within the framework of some models [121–123] in other cases. In particular, the hydrogen-like model by Rapp and Francis [121] with using the hydrogen wave functions in determination of the ion-atom exchange interaction potential is popular. But one can construct in this case a strict asymptotic theory by expansion of the cross section over a small parameter that exists in reality. In contrast to model evaluations, the asymptotic theory allows us to estimate the accuracy of the results within the theory accuracy because it is expressed through a small parameter of the theory. We show this below.

Note the classical character of nuclear motion at not small collision energies [124] (say, above those at room temperature). If an atom is found in a highly excited state, the electron transition in the resonant charge exchange process has a classical character also [124] and proceeds when a barrier between fields of two cores disappears. We consider another case related to atoms in the ground and lowest excited states, when the transition has a tunnel character. Then we have a weak logarithm dependence of the cross section of resonant charge exchange σ_{res} on the collision velocity v [125, 126]

$$\sigma_{res}(v) = C \ln^2 \left(\frac{v_*}{v} \right), \qquad (6.9)$$

where C and v_* are constants. Indeed, taking the basic dependence of the exchange interaction potential $\Delta(R)$ on a ion-atom distance R as $\Delta(R) \backsim \exp(-\gamma R)$ according to formula (5.10), one can represent the cross section of resonant charge exchange in the form

$$\sigma_{res}(v) = \frac{\pi}{2} \left[R_o + \frac{1}{\gamma} \ln \left(\frac{v_o}{v} \right) \right]^2, \qquad (6.10)$$

where $\pi R_o^2/2$ is the cross section at the collision velocity v_o. Diagram of Fig. 6.4 gives values of the cross sections of resonant charge exchange involving ions and atoms in the ground states for various elements. The collision energy at which the cross section is given corresponds to the laboratory frame of reference, where an atom is motionless.

We will be guided by collision energies of the order of eV, at which the parameter γR_o in formula (6.10) is large. The diagram of Fig. 6.5 gives values of the parameter

γR_o for ions and atoms of various elements. From this it follows that this parameter for various elements for the collision energy of the order of 1 eV ranges approximately from 10 up 15. On the basis of this one can construct a strict asymptotic theory of the resonant charge exchange by expansion the cross section of this process over a small parameter $1/\gamma R_o$. The first term of this expansion leads to the cross section of resonant charge exchange in the form [127]

$$\sigma_{res}(v) = \frac{\pi}{2} R_o^2, \quad \zeta(R_o) = \frac{e^{-C}}{2} = 0.28, \tag{6.11}$$

where $C = 0.577$ is the Eiler constant. Let us ignore elastic ion-atom scattering in the charge exchange process, i.e. particles move along straightforward trajectories, and the ion-atom distance R varies in time as $R^2 = \rho^2 = v^2 t^2$, where ρ is the impact parameter of collision, and v is the relative velocity of colliding particles. Then the cross section of resonant charge exchange is given by [128]

$$\sigma_{res}(v) = \frac{\pi}{2} R_o^2, \quad \frac{1}{v}\sqrt{\frac{\pi R_o}{2\gamma}} \Delta(R_o) = 0.28, \tag{6.12}$$

where the exchange interaction potential $\Delta(R)$ in this formula is determined by formula (5.8). Then the cross section (6.12) of resonant charge exchange is expressed through the parameters of the asymptotic electron wave function, when a valence electron is located in the atom far from the core.

Because this theory represents the cross section as an expansion over a small parameter $1/\gamma R_o$, one can estimate the accuracy of the expression (6.12) within the framework of the theory by comparison it with the results where the next terms of expansion over a small parameter are taken into account. In the hydrogen case (that is for the process $H^+ + H$) the cross section of resonant charge exchange at the ion energy of 1 eV is [129] $(173 \pm 2)a_o^2$ as averaging of results with accounting for the first and the second terms of expansion in different versions of this expansion. Evidently the accuracy from this $\sim 1\%$ is the best accuracy that one can expect from the asymptotic theory of resonant charge exchange described by formula (6.12) [129]. We estimate the accuracy of data for the cross section of resonant charge exchange represented in diagram of Fig. 6.5 to be better than 2% for elements of the first and second groups of the periodical table with the accuracy which is better, where ions and their atoms are found in the ground states and resonant charge exchange results from transition of s-electron.

Resonant charge exchange for ions and atoms of other groups of the periodical table of elements becomes more complicated because the process of resonant charge exchange is entangled with the processes of turning of atom angular momenta and transition between states of fine structure. In particular, Fig. 5.17 gives the lowest electron terms in the case of collision $Cl^+(^3P) + Cl(^2P)$ [100] which may be responsible for resonant charge exchange. Below we will be guided on the average cross section of resonant charge exchange that is averaged over initial values of atom and ion momenta and their projections. In particular, there are 18 electron even and odd terms in the case of interaction $Cl^+(^3P) + Cl(^2P)$ (Fig. 5.17) where the quantum numbers are the total momenta J and j of an atom and ion and M_J the atom momentum projection onto the impact parameter of collision. Let us consider the case when a valence p-electron is located in the fields of structureless cores. It is convenient to compare the parameters in this case with those for s-electron if the asymptotic radial wave functions (3.19) are identical in both cases. Then the exchange interaction potential of ion and atom with a valence p-electron is expressed through that Δ_o for s-electron given by formula (5.10) as [97]

$$\Delta_{lm}(R) = \Delta_o(R) \cdot \frac{(2l+1)(l+|m|)!}{(l-|m|!)|m|!}, \tag{6.13}$$

where l, m are the electron moment and its projection onto the molecular axis.

Table 6.1 gives the partial cross sections of resonant charge exchange with transition p-electron between two structureless cores, so that σ_o is the cross section with participation of s-electron, σ_{10} is the cross section involving p-electron if the momentum projection on the impact parameter of collision is zero, σ_{11} is this cross section when the momentum projection equals to ± 1, the average cross section is $\overline{\sigma} = \sigma_{10}/3 + 2\sigma_{11}/3$. These cross sections relate to the "a" case of Hund coupling where spin-orbit interaction is ignored. The cross sections $\sigma_{1/2}, \sigma_{3/2}$ correspond to case "c" of Hund coupling where spin-orbit splitting of levels is relatively large. As is seen, in the case "c" of Hund coupling the cross section is independent practically of the total electron momentum. In addition, the average cross sections are nearby for both cases of Hund coupling. Note that to this one-electron cases relate to elements of 3 and 8 groups of the periodical system of elements [61]. The accuracy of the cross sections given in diagram of Fig. 6.5 are determined by the accuracy of asymptotic coefficients A mostly and are estimated for these cases better than 5% at low collision energies.

The average cross section of resonant charge exchange for ions and atoms with noncompleted shells depends on the initial distributions over electron states of ions and atoms as well as on the character of momentum coupling in them as it follows from the analysis for halogens, oxygen and nitrogen [100, 130, 131]. Table 6.2 compares the average cross sections for resonant charge exchange involving atoms and their ions with valence p-electrons with those where atoms and ions contain valence s-electrons, but parameters A and γ are identical in these cases. The cross sections of resonant charge exchange for valence p-electrons relate to the case "a" of Hund coupling where spin-orbit interaction may be ignored. Note that transition to elements of 6, 7 and 8 groups of the periodical system from elements of 3, 4 and 5 groups results in transition from electrons to holes. The accuracy of the cross sections in diagram of Fig. 6.4 in these cases is estimated better than 10% [132]. As is seen, though the partial cross sections differ remarkable, the average cross sections are nearby for different types of momentum coupling.

Representing the velocity dependence for the cross section of resonant charge exchange in the form

$$\sigma_{res}(v) = \sigma_{res}(v_o)\left(\frac{v_o}{v}\right)^{\alpha},\tag{6.14}$$

we obtain from formula (6.10)

$$\alpha = \frac{2}{\gamma R_o} \ll 1$$

In reality, the exponent α differs from this value because of an approximated character of formula (6.10). The values of this exponent for the resonant charge exchange process involving various elements are given in diagram of Fig. 6.5.

6.3 Elastic Collisions of Electrons with Atoms

Collisions of electrons with atomic particles in gases and plasmas determine some transport properties of these systems and are an important object for the analysis. The peculiarity of scattering of electrons on atomic particles is such that numerical evaluations of the cross section of such processes at not large collision velocities are non-reliable because of the sensitivity of results to weak exchange interactions of an incident electron with internal atomic electrons. Therefore the reliable cross sections of electron-atom collisions at not large electron energies follow from the experiment only, whereas the theory gives general relations between parameters of these processes. Basing on the theoretical analysis, we below consider a general picture for electron-atom scattering, while numerical parameters of this picture follows from the experiment.

Electron scattering on atomic particles in gases and plasmas has a quantum character. Assuming the effective electron-atom potential to be spherically symmetric, one can consider scattering of an electron on a motionless atom to be independently for different electron momenta l. This is a basis of the partial wave method [136], where the scattering parameters are the sum of these partial parameters at certain l. The characteristic of electron-atom scattering for a given l is the scattering phase δ_l, that describes the asymptotic wave function of the electron-atom system at large distances between them. Formulas for the differential cross section $d\sigma = 2\pi|f(\vartheta)|^2 d\cos\vartheta$ of electron-atom scattering, the scattering amplitude $f(\vartheta)$, the diffusion cross section σ^* and the scattering phases δ_l have the form [46, 136]

$$f(\vartheta) = \frac{1}{2iq}\sum_{l=0}^{\infty}(2l+1)(e^{2i\delta_l}-1)P_l(\cos\vartheta), \quad \sigma^* = \frac{4\pi}{q^2}\sum_{l=0}^{\infty}(l+1)\sin^2(\delta_l-\delta_{l+1}),$$

$$\tag{6.15}$$

where q is the electron wave vector, so that the electron energy ε is expressed through the electron wave vector by the relation $\varepsilon = \hbar^2 q^2/2m_e$ (m_e is the electron mass). We account for the electron mass to be small in comparison with the atom mass, so that the reduced mass of colliding particles coincides practically with the electron mass.

Note that the partial wave method is suitable for electron scattering on a structureless atom.

In the limit of small values of the wave vector q, scattering parameters (6.15) are expressed through the zero phase only, that is given by $\delta_o = -Lq$ in this limit, where L is the scattering length. Correspondingly, the scattering parameters (6.15) are equal in this limit of small q

$$f(\vartheta) = -L, \ \sigma^*(0) = 4\pi L^2 \tag{6.16}$$

If we approximate the effective electron-atom interaction potential as a short-range one, the effective interaction potential $U(\mathbf{r})$ is expressed through the scattering length L by formula [90]

$$U_{sh}(\mathbf{r}) = 2\pi L \frac{\hbar^2}{m_e} \delta(\mathbf{r}), \tag{6.17}$$

if the electron wavelength exceeds a size of the atom potential well. The scattering length coincides with the atom effective radius [137], if the assumption is used that an electron cannot penetrate inside a region that is restricted by the effective radius. In addition, Table 6.3 contains values of the scattering length for an electron on inert gas atoms.

Note that parameters of electron-atom scattering are introduced through the behavior of the wave function of a scattering electron where an interaction between the incident electron and valence electrons of an atom becomes weak. One can define the electron-atom scattering length assuming that interaction takes place in the atom region which size is small compared with the wavelength of an incident electron. Then the definition of the electron-atom scattering length L follows from the following boundary condition for the wave function Ψ of the scattered electron

$$\frac{\mathrm{d}\ln\Psi}{\mathrm{d}r}\Big|_{r=0} = -\frac{1}{L}, \tag{6.18}$$

where r is a distance of the scattering electron from the atom center. In this definition, the scattering length is determined by electron-atom interaction inside the atom, where a one-electron approximation is not correct, i.e. the wave function of a scattering electron is entangled with the wave functions of atomic electrons, and a resultant exchange interaction has a complex form.

If the electron-atom scattering length is negative, the zero scattering phase δ_o becomes zero at a low electron energy, whereas other scattering phases δ_l are small. As a result, the electron-atom cross section (both diffusion, and total) acquires a deep minimum at low electron energies that is known as the Ramsauer effect [141, 142]. According to the data of Table 6.3, the Ramsauer effect is realized in the case of electron scattering on argon, krypton and xenon atoms. Figure 6.6 gives experimental values of the diffusion cross section of electron scattering on a xenon atoms based on measurements [143–147], and Fig. 6.7 represents the diffusion cross sections of electron scattering on inert gas atoms which result from the sum of measurements [138].

In the case where the electron-atom interaction potential is the sum (6.17) of a short-range interaction potential and the polarization interaction potential $U_l(r) = -\alpha e^2/2r^4$ as a long-range interaction potential, these interactions may be divided at low electron energies, and the the scattering phases δ_l may be represented in the form of expansion over a small electron wave vector q [148, 149]. This expansion is based on a small parameter qa/La_o (a_o is the Bohr radius).

In this limit the electron scattering phase is $\delta_o = -Lq$, and this may be also a definition of the scattering length L. Correspondingly, the cross sections of electron-atom scattering at small electron energies are equal to

$$\sigma^*(0) = \sigma_t(0) = 4\pi L^2 \tag{6.19}$$

Other scattering phases have a more strong dependence on the wave vector in the limit of its low values. In the case of the polarization long-range interaction $U(r) = -\alpha e^2/2r^4$ at an electron distance r from an atom and a short-range interaction in the form (5.11), the scattering amplitude at low electron energies is [150]

$$f(\vartheta) = -L - \frac{\pi\alpha q}{2a_o} \sin\frac{\vartheta}{2} \tag{6.20}$$

This expansion leads to the following expression for the diffusion cross section of electron-atom scattering [150, 151]

$$\sigma^* = 4\pi L^2 \left(1 - \frac{8}{5}x + \frac{2}{3}x^2\right), \quad x = -\frac{\pi\alpha q}{2La_o} \tag{6.21}$$

This cross section has a minimum if the wave vector is $q_{min} = -12La_o/(5\pi\alpha)$ ($x = 6/5$) and the minimal cross section is 25 times lower than that at zero electron energy. Though this analysis with using a short-range and polarization electron-atom interactions exhibits a deep minimum in the diffusion electron-atom cross section at a negative electron-atom scattering length, in reality the minimal cross section lower compared to this and differs from the cross section at zero electron energy by approximately two orders of magnitude. Table 6.3 contains the energies ε_{min} at which the minimum cross section of elastic electron-atom scattering is observed for inert gas atoms, as well as values of the minimal cross sections $\sigma_t(\varepsilon_{min})$ at such energies.

The above analysis relates to electron scattering on a structureless atom. In the case of electron scattering on atoms with noncompleted electron shells, different channels of electron scattering are present in the scattering amplitude. Let us consider as an example electron scattering on an alkali metal atom in the ground state, where the total electron-atom system may have a spin $S = 0; 1$, and scattering proceeds independently in each channel, so that at zero electron energy the diffusion cross section of electron atom scattering is equal to

$$\sigma^*(0) = \sigma_t(0) = \pi(L_o^2 + 3L_1^2), \tag{6.22}$$

where L_o and L_1 are the electron scattering lengths if the total spin of the electron-atom system is zero and one correspondingly. These scattering lengths are given in Table 6.4. Note the importance of the resonance ^3P in electron scattering on alkali metal atoms. Parameters of this resonance, the excitation energy of this autodetaching state E_r and its width Γ_r, are given in Table 6.4.

6.4 Elementary Processes in Atmospheric Plasma

Processes in excited and ionized gases are determined a current state of an atomic system and its evolution. Because of a large variety of these processes, various physical situations are possible for a state and development of an atomic system. As a demonstration of a variety of such processes, we represent below rates of processes which partake in an atmospheric plasma [33, 135, 152, 153] which consists of atomic particles of nitrogen and oxygen only. Table 6.5 contains the rate constants of processes where electrons are generated or are lost. Since ions are the basic charged atmospheric component, the processes involving ions are of importance for electric phenomena in the Earth's atmosphere. The rate constants of processes of ion chemistry in the atmosphere are given in Table 6.6. These processes include both positive and negative atmospheric ions. The rate constants of chemical processes in atmospheric air are given in Table 6.7. Table 6.8 represents values of the rate constants for three-body processes at room temperature with participation of atmospheric ions.

Figures

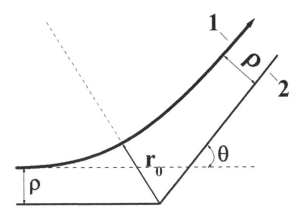

Fig. 6.1 Parameters of classical scattering in the center-of-mass frame of reference for colliding particles. An arrow indicates the direction of the particle trajectory in the center-of-mass frame of reference, ρ is the impact parameter of collision, r_o is the distance of closest approach, ϑ is the scattering angle

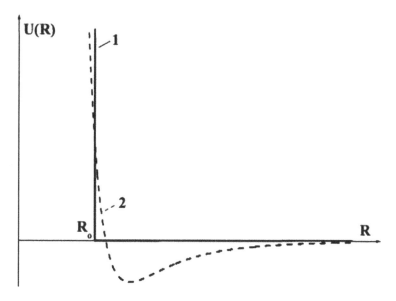

Fig. 6.2 1—the interaction potential of atomic particles for the hard sphere model, 2—real interaction potential of atomic particles

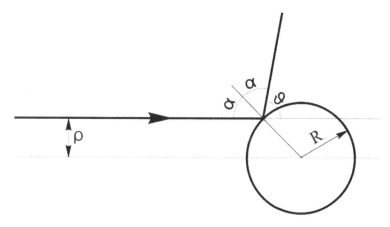

Fig. 6.3 The character of scattering within the framework of the hard sphere model

Cross sections of reso

Group / Period	I	II	III	IV	V
1	1.008 1s $^2S_{1/2}$ $_1$H 1.000 6.12 Hydrogen 2.00 4.82 3.65	1s² $^1S_0$ 4.003 3.4 1.344 $_2$He 2.7 2.87 Helium 2.1			
2	6.491 2s $^2S_{1/2}$ Li 0.630 26 Lithium 0.82 22 18	9.012 3s² $^1S_0$ $_4$Be 0.828 13 Berillium 1.6 10 8.3	2p $^2P_{1/2}$ 10.81 14 0.781 $_5$B 12 0.88 Boron 9.0	2p² $^3P_0$ 12.011 9.9 0.910 $_6$C 8.2 1.3 Carbon 6.6	2p³ $^4S_{3/2}$ 14.007 7.7 1.034 $_7$N 6.4 1.5 Nitrogen 5.2
3	22.990 3s $^2S_{1/2}$ $_{11}$Na 0.615 31 Sodium 0.74 26 21	24.305 3s² $^1S_0$ $_{12}$Mg 0.750 18 1.3 15 Magnesium 12	3p $^2P_{1/2}$ 26.982 21 0.663 $_{13}$Al 19 0.61 17 Aluminium	3p² $^3P_0$ 28.086 17 0.774 $_{14}$Si 14 1.1 Silicon 12	3p³ $^4S_{3/2}$ 30.974 14 0.878 $_{15}$P 12 1.6 10 Phosphorus
4	39.098 4s $^2S_{1/2}$ $_{19}$K 0.565 40 Potassium 0.52 34 28	40.08 4s² $^1S_0$ $_{20}$Ca 0.670 25 Calsium 0.95 21 17	44.956 3d4s² $^2D_{3/2}$ $_{21}$Sc 0.693 24 1.1 20 Scandium 16	47.88 3d²4s² $^3F_2$ $_{22}$Ti 0.708 22 1.2 19 Titanium 15	50.942 3d³4s² $^4F_{3/2}$ $_{23}$V 0.704 23 1.2 19 Vanadium 16
	3d¹⁰4s $^2S_{1/2}$ 63.546 19 0754 $_{29}$Cu 16 1.3 13 Copper	4s² $^1S_0$ 65.38 15 0.831 $_{30}$Zn 12 1.7 10 Zinc	4p $^2P_{1/2}$ 69.72 26 0.664 $_{31}$Ga 22 0.60 18 Gallium	4p² $^3P_0$ 72.59 20 0.762 $_{32}$Ge 17 1.3 14 Germanium	4p³ $^4S_{3/2}$ 74.922 15 0.850 $_{33}$As 13 1.6 11 Arsenic
5	85.468 5s $^2S_{1/2}$ $_{37}$Rb 0.554 45 Rubidium 0.48 38 32	87.62 5s² $^1S_0$ $_{38}$Sr 0.647 29 0.86 25 Strontium 20	88.906 4d5s² $^2D_{3/2}$ $_{39}$Y 0.682 25 1.0 18 Yttrium	91.22 4d²5s² $^3F_2$ $_{40}$Zr 0.709 23 1.2 20 Zirconium 16	92.906 4d⁴5s $^6D_{1/2}$ $_{41}$Nb 0.711 23 1.2 19 Niobium 16
	4d¹⁰5s $^2S_{1/2}$ 107.87 20 0.746 $_{47}$Ag 17 1.2 14 Silver	5s² $^1S_0$ 112.41 16 0.813 $_{48}$Cd 14 1.6 11 Cadmium	5p $^2P_{1/2}$ 114.82 28 0.652 $_{49}$In 24 0.58 20 Indium	5p² $^3P_0$ 118.69 22 0.735 $_{50}$Sn 19 1.0 15 Tin	5p³ $^4S_{3/2}$ 121.75 20 0.797 $_{51}$Sb 17 1.7 14 Antimony
6	132.90 6s $^2S_{1/2}$ $_{55}$Cs 0.535 51 Cesium 0.41 44 36	137.33 6s² $^1S_0$ $_{56}$Ba 0.619 35 0.78 30 Barium 25	138.90 5d6s² $^2D_{3/2}$ $_{57}$La 0.640 32 0.90 27 Lanthanum 23	178.49 5d²6s² $^3F_2$ $_{72}$Hf 0.740 22 1.3 18 Hafnium 15	180.95 5d³6s² $^4F_{3/2}$ $_{73}$Ta 0.762 20 1.4 17 Tantalum 14
	5d¹⁰6s $^2S_{1/2}$ 196.97 16 0.823 $_{79}$Au 14 1.6 11 Gold	5d¹⁰6s² $^1S_0$ 200.59 14 0.876 $_{80}$Hg 12 1.9 10 Mercury	6p $^2P_{1/2}$ 204.38 26 0.670 $_{81}$Tl 22 0.55 18 Thallium	6p² $^3P_0$ 207.2 22 0.738 $_{82}$Pb 19 1.1 16 Lead	6p³ $^4S_{3/2}$ 208.98 26 0.732 $_{83}$Bi 22 1.4 19 Bismuth
7	$[223]$ 7s $^2S_{1/2}$ $_{87}$Fr 0.542 53 0.49 45 Francium 38	226.02 7s² $^1S_0$ $_{88}$Ra 0.623 35 0.78 30 Radium 25	227.03 6d7s² $^2D_{3/2}$ $_{89}$Ac 0.636 33 0.70 28 Actinium 24		

Actinides.

6d²7s² $^3F_2$ 232.04 0.675 28 $_{90}$Th 0.94 24 Thorium 20	5f²6d7s² $^4K_{11/2}$ 231.04 0.647 30 $_{91}$Pa 0.70 25 Protactinium 21	5f³6d7s² $^5L_6$ 238.03 0.670 28 $_{92}$U 0.89 24 Uranium 20	5f⁴6d7s² $^6L_{11/2}$ 237.05 0.580 46 $_{93}$Np 0.70 39 Neptunium 33	$[244]$ 5f⁶7s² $^7F_0$ 0.612 38 $_{94}$Pu 0.73 32 Plutonium 27	5f⁷7s² $^8S_{7/2}$ $[243]$ 0.569 49 $_{95}$Am 0.70 42 Americium 36

Fig. 6.4 Cross section of resonant charge exchange

nant charge exchange.

Legend:

Shell of valence electrons — Electron term

Atomic weight: *137.33* $6s^2\,{}^1S_0$

Symbol: **Ba**

Atomic number: 56   0.619   35

Element: Barium   0.78   30   25

Asymptotic parameters

Cross section of resonant charge exchange (in $10^{-15}\,cm^2$ at 0.1, 1, 10 eV in the laboratory frame)

Group VI

Element	Atomic weight	Valence shell / term	Asymptotic params	Cross section	Second value	σ at 0.1/1/10 eV
$_8$O Oxygen	15.999	$2p^4\,{}^3P_2$	8.3 / 6.9 / 5.6	1.000	1.3	—
$_{16}$S Sulfur	32.06	$3p^4\,{}^3P_2$	13 / 10 / 8.6	0.873	1.1	—
$_{24}$Cr Chromium	51.996	$3d^5 4s\,{}^7S_3$	—	0.705	1.1	23 / 19 / 16
$_{34}$Se Selenium	78.96	$4p^4\,{}^3P_2$	16 / 14 / 12	0.847	1.5	—
$_{42}$Mo Molibdenum	95.94	$4d^5 5s\,{}^7S_3$	1.2	0.722	—	22 / 19 / 15
$_{52}$Te Tellurium	127.60	$5p^4\,{}^3P_2$	18 / 16 / 13	0.814	1.6	—
$_{74}$W Tungsten	183.85	$5d^4 6s^2\,{}^5D_0$	1.4	0.766	—	20 / 17 / 14
$_{84}$Po Polonium	[209]	$6p^4\,{}^3P_2$	17 / 15 / 12	0.788	1.5	—

Group VII

Element	Atomic weight	Valence shell / term	Asymptotic params	Cross section	Second value	σ at 0.1/1/10 eV
$_9$F Fluorine	18.998	$2p^5\,{}^2P_{3/2}$	5.7 / 4.7 / 3.7	1.132	1.6	—
$_{17}$Cl Chlorine	35.453	$3p^5\,{}^2P_{3/2}$	9.6 / 8.0 / 6.6	0.976	1.8	—
$_{25}$Mn Manganese	54.938	$3d^5 4s^2\,{}^6S_{5/2}$	1.3	0.739	—	20 / 17 / 14
$_{35}$Br Bromine	79.904	$4p^5\,{}^2P_{3/2}$	12 / 10 / 8.2	0.932	1.8	—
$_{43}$Tc Technetium	[98]	$4d^5 5s^2\,{}^6S_{5/2}$	1.3	0.731	—	21 / 18 / 15
$_{53}$I Iodine	126.90	$5p^5\,{}^2P_{3/2}$	14 / 12 / 10	0.876	1.9	—
$_{75}$Re Rhenium	186.21	$5d^5 6s^2\,{}^6S_{5/2}$	1.4	0.761	—	20 / 17 / 14
$_{85}$At Astatine	[210]	$6p^5\,{}^2P_{3/2}$	18 / 15 / 13	0.813	1.9	—

Group VIII

Element	Atomic weight	Valence shell / term	Asymptotic params	Cross section	Second value	σ at 0.1/1/10 eV
$_{10}$Ne Neon	20.179	$2p^6\,{}^1S_0$	4.1 / 3.2 / 2.6	1.259	1.8	—
$_{18}$Ar Argon	39.948	$2p^6\,{}^1S_0$	7.0 / 5.8 / 4.7	1.076	2.0	—
$_{26}$Fe Iron	55.847	$3d^6 4s^2\,{}^5D_4$	1.4	0.762	—	19 / 16 / 13
$_{36}$Kr Krypton	83.80	$4p^6\,{}^1S_0$	9.0 / 7.5 / 6.2	1.014	2.1	—
$_{44}$Ru Ruthenium	101.07	$4d^7 5s\,{}^5F_5$	1.2	0.736	—	18 / 18 / 15
$_{54}$Xe Xenon	131.29	$5p^6\,{}^1S_0$	12 / 10 / 8.6	0.944	2.2	—
$_{76}$Os Osmium	190.2	$5d^6 6s^2\,{}^5D_4$	1.7	0.801	—	18 / 15 / 13
$_{86}$Rn Radon	[222]	$6p^6\,{}^1S_0$	15 / 13 / 11	0.889	2.3	—

Group VIII (triad elements)

Element	Atomic weight	Valence shell / term	Cross section	Second value	σ at 0.1/1/10 eV
$_{27}$Co Cobalt	58.933	$3d^7 4s^2\,{}^4F_{9/2}$	0.760	1.4	19 / 16 / 13
$_{28}$Ni Nickel	58.69	$3d^8 4s^2\,{}^3F_4$	0.749	1.4	20 / 17 / 14
$_{45}$Rh Rhodium	102.91	$4d^8 5s\,{}^4F_{9/2}$	0.741	1.2	20 / 17 / 14
$_{46}$Pd Palladium	106.42	$4d^{10}\,{}^1S_0$	0.783	2.1	18 / 16 / 13
$_{77}$Ir Iridium	192.22	$5d^7 6s^2\,{}^4F_{9/2}$	0.816	1.7	17 / 14 / 12
$_{78}$Pt Platinum	195.08	$5d^9 6s\,{}^3D_3$	0.812	1.5	17 / 14 / 12

Lantanides.

Element	Atomic weight	Valence shell / term	Cross section	Second value	σ at 0.1/1/10 eV
$_{58}$Ce Cerium	140.12	$4f 5d 6s^2\,{}^1G_4$	0.638	0.88	32 / 27 / 23
$_{59}$Pr Praseodymium	140.91	$4f^3 6s^2\,{}^4I_{9/2}$	0.634	0.84	32 / 28 / 23
$_{60}$Nd Neodymium	144.24	$4f^4 6s^2\,{}^5I_4$	0.637	0.85	32 / 27 / 23
$_{61}$Pm Promethium	[145]	$4f^5 6s^2\,{}^6H_{5/2}$	0.640	0.86	32 / 27 / 22
$_{62}$Sm Samarium	150.36	$4f^6 6s^2\,{}^7F_0$	0.644	0.88	31 / 26 / 22
$_{63}$Eu Europium	151.96	$4f^7 6s^2\,{}^8S_{7/2}$	0.646	0.89	31 / 26 / 22
$_{64}$Gd Gadolinium	157.25	$4f^7 5d 6s^2\,{}^9D_2$	0.672	1.0	28 / 24 / 20
$_{65}$Tb Terbium	158.92	$4f^9 6s^2\,{}^6H_{15/2}$	0.657	0.93	30 / 25 / 21
$_{66}$Dy Dysprosium	162.50	$4f^{10} 6s^2\,{}^5I_8$	0.661	0.94	29 / 25 / 20
$_{67}$Ho Holmium	164.93	$4f^{11} 6s^2\,{}^5I_{15/2}$	0.665	0.96	29 / 24 / 20
$_{68}$Er Erbium	167.26	$4f^{12} 6s^2\,{}^3H_6$	0.670	0.98	28 / 24 / 20
$_{69}$Tm Thulium	168.93	$4f^{13} 6s^2\,{}^2F_{7/2}$	0.674	0.99	28 / 23 / 19
$_{70}$Yb Ytterbium	173.04	$4f^{14} 6s^2\,{}^1S_0$	0.678	1.0	27 / 23 / 19
$_{71}$Lu Lutetium	174.97	$4f^{14} 5d 6s^2\,{}^2D_{3/2}$	0.632	0.92	34 / 29 / 24

Fig. 6.4 (continued)

Group \\ Period	I	II	Parameters of cross section of		
			III	**IV**	**V**
1	1.008 1s $^2S_{1/2}$ $_1$H 1.000 4.82 Hydrogen 10 0.22	1s² ¹S₀ 4.003 2.7 1.344 $_2$He 0.22 10 Helium			
2	6.491 2s $^2S_{1/2}$ $_3$Li 0.630 22 Lithium 14 0.16	9.012 2s² ¹S₀ $_4$Be 0.828 10 Beryllium 12 0.20	2p $^2P_{1/2}$ 10.81 12 0.781 $_5$B 0.19 13 Boron	2p² ³P₀ 12.011 8.2 0.910 $_6$C 12 0.18 Carbon	2p³ $^4S_{3/2}$ 14.007 6.4 1.034 $_7$N 12 0.17 Nitrogen
3	22.990 3s $^2S_{1/2}$ $_{11}$Na 0.615 26 Sodium 15 0.17	24.305 3s² ¹S₀ $_{12}$Mg 0.750 15 Magnesium 14 0.18	3p $^2P_{1/2}$ 26.982 19 0.663 $_{13}$Al 0.12 14 Aluminium	3p² ³P₀ 28.086 14 0.774 $_{14}$Si 0.15 14 Silicon	3p³ $^4S_{3/2}$ 30.974 12 0.878 $_{15}$P 0.15 14 Phosphorus
4	39.098 4s $^2S_{1/2}$ $_{19}$K 0.565 34 Potassium 16 0.16	40.08 4s² ¹S₀ $_{20}$Ca 0.670 21 Calcium 15 0.17	44.956 3d4s² $^2D_{3/2}$ $_{21}$Sc 0.693 20 Scandium 15 0.18	47.88 3d²4s² $^3F_2$ $_{22}$Ti 0.708 19 Titanium 15 0.17	50.942 3d³4s² $^4F_{3/2}$ $_{23}$V 0.704 19 15 Vanadium 0.16
	3d¹⁰4s $^2S_{1/2}$ 63.546 16 0.754 $_{29}$Cu 0.16 14 Copper	4s² ¹S₀ 65.38 12 0.831 $_{30}$Zn 0.18 14 Zinc	4p $^2P_{1/2}$ 69.72 22 0.664 $_{31}$Ga 0.16 15 Gallium	4p² ³P₀ 72.59 17 0.762 $_{32}$Ge 0.16 15 Germanium	4p³ $^4S_{3/2}$ 74.922 13 0.850 $_{33}$As 0.14 15 Arsenic
5	85.468 5s $^2S_{1/2}$ $_{37}$Rb 0.554 38 Rubidium 16 0.15	87.62 5s² ¹S₀ $_{38}$Sr 0.647 25 Strontium 15 0.16	88.906 4d5s² $^2D_{3/2}$ $_{39}$Y 0.682 21 Yttrium 15 0.14	91.22 4d²5s² $^3F_2$ $_{40}$Zr 0.709 20 Zirconium 15 0.16	92.906 4d⁴5s $^6D_{1/2}$ $_{41}$Nb 0.711 19 Niobium 15 0.16
	4d¹⁰5s $^2S_{1/2}$ 107.87 17 0.746 $_{47}$Ag 0.16 15 Silver	5s² ¹S₀ 112.41 14 0.813 $_{48}$Cd 0.16 14 Cadmium	5p $^2P_{1/2}$ 114.82 24 0.652 $_{49}$In 0.15 15 Indium	5p² ³P₀ 118.69 19 0.735 $_{50}$Sn 0.17 15 Tin	5p³ $^4S_{3/2}$ 121.75 17 0.797 $_{51}$Sb 0.16 16 Antimony
6	132.90 6s $^2S_{1/2}$ $_{55}$Cs 0.535 44 Cesium 17 0.15	137.33 6s² ¹S₀ $_{56}$Ba 0.619 30 Barium 16 0.15	138.90 5d6s² $^2D_{3/2}$ $_{57}$La 0.640 27 Lanthanum 16 0.14	178.49 5d²6s² $^3F_2$ $_{72}$Hf 0.740 18 Hafnium 15 0.13	180.95 5d³6s² $^4F_{3/2}$ $_{73}$Ta 0.762 17 Tantalum 15 0.16
	5d¹⁰6s $^2S_{1/2}$ 196.97 14 0.823 $_{79}$Au 0.16 15 Gold	5d¹⁰6s² ¹S₀ 200.59 12 0.876 $_{80}$Hg 0.15 14 Mercury	6p $^2P_{1/2}$ 204.38 22 0.670 $_{81}$Tl 0.16 15 Thallium	6p² ³P₀ 207.2 19 0.738 $_{82}$Pb 0.14 15 Lead	6p³ $^4S_{3/2}$ 208.98 22 0.732 $_{83}$Bi 0.14 16 Bismuth
7	[223] 7s $^2S_{1/2}$ $_{87}$Fr 0.542 45 Francium 17 0.14	226.02 7s² ¹S₀ $_{88}$Ra 0.623 30 Radium 16 0.15	227.03 6d7s² $^2D_{3/2}$ $_{89}$Ac 0.636 28 Actinium 16 0.14		

Fig. 6.5 Parameters of cross section of resonant charge exchange

of resonant charge exchange.

Legend:

Shell of valence electrons
γ \ Electron term

Atomic weight — 24.305 $3s^2\ {}^1S_0$ Cross section of resonant charge exchange in $10^{-15}\,\text{cm}^2$ at 1 eV in the laboratory frame

Symbol — 0.750 15
Atomic number — $_{12}$Mg 14
Element — Magnesium 0.18

$\gamma R_0\ \alpha$

VI	VII	VIII		
$2p^4\,{}^3P_2$ 15.999 6.9 1.000 $_8$O 12 0.17 Oxygen	$2p^5\,{}^2P_{3/2}$ 18.998 4.7 1.132 $_9$F 12 0.19 Fluorine	$2p^6\,{}^1S_0$ 20.179 3.2 1.259 $_{10}$Ne 0.20 11 Neon		
$3p^4\,{}^3P_2$ 32.06 10 0.873 $_{16}$S 0.18 13 Sulfur	$3p^5\,{}^2P_{3/2}$ 35.453 8.0 0.976 $_{17}$Cl 0.16 13 Chlorine	$2p^6\,{}^1S_0$ 39.948 5.8 1.076 $_{18}$Ar 0.17 12 Argon		
51.996 $3d^54s\,{}^7S_3$ $_{24}$Cr 0.705 19 Chromium 15 0.16	54.938 $3d^54s^2\,{}^6S_{5/2}$ $_{25}$Mn 0.739 17 Manganese 14 0.16	55.847 $3d^64s^2\,{}^5D_4$ $_{26}$Fe 0.762 16 Iron 14 0.16	58.933 $3d^74s^2\,{}^4F_{9/2}$ $_{27}$Co 0.760 16 Cobalt 14 0.16	58.69 $3d^84s^2\,{}^3F_4$ $_{28}$Ni 0.749 16 Nickel 15 0.16
$4p^4\,{}^3P_2$ 78.96 14 0.847 $_{34}$Se 0.13 15 Selenium	$4p^5\,{}^2P_{3/2}$ 79.904 10 0.932 $_{35}$Br 0.16 14 Bromine	$4p^6\,{}^1S_0$ 83.80 7.5 1.014 $_{36}$Kr 0.16 13 Krypton		
95.94 $4d^55s\,{}^7S_3$ $_{42}$Mo 0.722 19 Molibdenum 15 0.17	$[98]$ $4d^55s^2\,{}^6S_{5/2}$ $_{43}$Tc 0.731 18 Technetium 15 0.18	101.07 $4d^75s\,{}^5F_5$ $_{44}$Ru 0.736 18 Ruthenium 15 0.18	102.91 $4d^85s\,{}^4F_{9/2}$ $_{45}$Rh 0.741 17 Rhodium 15 0.16	106.42 $4d^{10}\,{}^1S_0$ $_{46}$Pd 0.783 16 Palladium 15 0.14
$5p^4\,{}^3P_2$ 127.60 16 0.814 $_{52}$Te 0.14 16 Tellurium	$5p^5\,{}^2P_{3/2}$ 126.90 12 0.876 $_{53}$I 0.15 14 Iodine	$5p^6\,{}^1S_0$ 131.29 10 0.944 $_{54}$Xe 0.14 14 Xenon		
183.85 $5d^46s^2\,{}^5D_0$ $_{74}$W 0.766 17 Tungsten 15 0.16	186.21 $5d^56s^2\,{}^6S_{5/2}$ $_{75}$Re 0.761 17 Rhenium 15 0.16	190.2 $5d^66s^2\,{}^5D_4$ $_{76}$Os 0.801 15 Osmium 15 0.14	192.22 $5d^76s^2\,{}^4F_{9/2}$ $_{77}$Ir 0.816 14 Iridium 15 0.15	195.08 $5d^96s\,{}^3D_3$ $_{78}$Pt 0.812 14 Platinum 14 0.15
$6p^4\,{}^3P_2$ $[209]$ 15 0.788 $_{84}$Po 0.15 15 Polonium	$6p^5\,{}^2P_{3/2}$ $[210]$ 15 0.813 $_{85}$At 0.14 15 Astatine	$6p^6\,{}^1S_0$ $[222]$ 13 0.889 $_{86}$Rn 0.14 15 Radon		

Fig. 6.5 (continued)

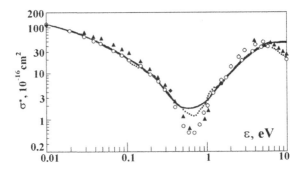

Fig. 6.6 The diffusion cross section of electron scattering on the xenon atom [143–147]

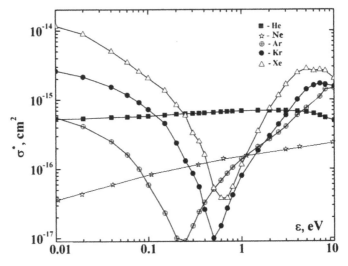

Fig. 6.7 Diffusion cross section of electron scattering on inert gas atoms depending on the electron energy [138–140] according to measurements

Tables

Table 6.1 The partial cross sections of resonant charge exchange with transition of p-electron between two structureless cores [61]

$R_o\gamma$	6	8	10	12	14	16
σ_{10}/σ_o	1.40	1.29	1.23	1.19	1.16	1.14
σ_{11}/σ_o	1.08	0.98	0.94	0.95	0.912	0.91
$\overline{\sigma}/\sigma_o$	1.19	1.08	1.04	1.01	0.99	0.99
$\sigma_{1/2}/\sigma_o$	1.18	1.10	1.07	1.05	1.04	1.03
$\sigma_{3/2}/\sigma_o$	1.18	1.10	1.06	1.04	1.03	1.02

Table 6.2 The reduced average cross sections of resonant charge exchange for atoms of 3 (σ_3), 4 (σ_4) and 5 (σ_5) groups of the periodical system of elements in the case "a" of Hund coupling [61, 133]

$R_o\gamma$	6	8	10	12	14	16
σ_3/σ_o	1.17	1.09	1.05	1.03	1.02	1.03
σ_4/σ_o	1.50	1.32	1.23	1.18	1.14	1.12
σ_5/σ_o	1.44	1.29	1.22	1.17	1.14	1.11

Table 6.3 Parameters of the total cross section σ_t for electron scattering on inert gas atoms [50, 138]

Atom	He	Ne	Ar	Kr	Xe
L/a_o	1.2	0.2	−1.6	−3.5	−6.5
$\sigma_t(\varepsilon = 0)$, Å^2	5.1	0.14	9.0	43	150
ε_{min}, eV	–	–	0.18	0.32	0.44
$\sigma_t(\varepsilon_{min})$, Å^2	–	–	0.88	3.8	13

Table 6.4 Electron scattering lengths on alkali metal atoms L_o and L_1, and the parameters of the autodetaching state ^3P for the negative alkali metal ion with the spin of one [50]

	Li	Na	K	Rb	Cs
L_o/a_o	3.6	4.2	0.56	2.0	−2.2
L_1/a_o	−5.7	−5.9	−15	−17	−24
E_r, eV	0.06	0.08	0.02	0.03	0.011
Γ_r, eV	0.07	0.08	0.02	0.03	0.008

Table 6.5 Rates of collision processes for formation and loss of electrons in atmospheric air. The rate constants k of pair processes are given in cm³/s and correspond to room temperature if the temperature or energy are not indicated

Type	Number	Process	Rate constant, cm³/s
Dissociative recombination	1	$e + N_2^+ \rightarrow N + N$	$2 \cdot 10^{-7}$
	2	$e + O_2 \rightarrow O + O$	$2 \cdot 10^{-7}$
	3	$e + NO^+ \rightarrow N + O$	$4 \cdot 10^{-7}$
	4	$e + N_4^+ \rightarrow N_2 + N_2^+$	$4 \cdot 10^{-7}$
	5	$e + O_4^+ \rightarrow O_2 + O_2^+$	$2.3 \cdot 10^{-6}$, T$=205$ K
Associative detachment of negative ions	6	$O^- + O \rightarrow e + O_2 + 3.6\,eV$	$2 \cdot 10^{-10}$
	7	$O^- + N \rightarrow e + NO + 5.1\,eV$	$3 \cdot 10^{-10}$
	8	$O^- + N_2 \rightarrow e + N_2O + 0.2\,eV$	$1 \cdot 10^{-14}$
	9	$O^- + O_2(^1\Delta_g) \rightarrow$ $e + O_3 + 0.5\,eV$	$3 \cdot 10^{-10}$
	10	$O^- + NO \rightarrow e + NO_2 + 1.6\,eV$	$1.4 \cdot 10^{-11}$
	11	$O^- + O_2 \rightarrow e + O_3 - 0.4\,eV$	$k < 1 \cdot 10^{-12}$
	12	$O^- + O_3 \rightarrow e + 2O_2 + 2.8\,eV$	$3 \cdot 10^{-10}$
	13	$O_2^- + O \rightarrow e + O_3 + 0.6\,eV$	$3 \cdot 10^{-10}$
	14	$O_2^- + N \rightarrow e + NO_2 + 4.1\,eV$	$5 \cdot 10^{-10}$
	15	$O_3^- + O \rightarrow e + 2O_2$	$1 \cdot 10^{-10}$
	16	$OH^- + O \rightarrow e + H_2O + 0.9\,eV$	$2 \cdot 10^{-10}$
	17	$OH^- + N \rightarrow e + HNO + 2.4\,eV$	$k < 1 \cdot 10^{-11}$
Electron release	18	$O_2^- + O_2 \rightarrow e + 2O_2 - 0.4\,eV$	$2 \cdot 10^{-18}$
	19	$O_2^- + O_2(^1\Delta_g) \rightarrow$ $e + 2O_2 + 0.6\,eV$	$2 \cdot 10^{-10}$
	20	$O_2^- + N_2 \rightarrow e + O_2 + O_2 - 0.4\,eV$	$2 \cdot 10^{-16}$
	21	$NO^- + NO \rightarrow e + 2NO$	$6 \cdot 10^{-12}$
Electron attachment	22	$e + 2O_2 \rightarrow O_2^- + O_2$	$K = 3 \cdot 10^{-30}$ cm⁶/s
	23	$e + O_2 + N_2 \rightarrow O_2^- + N_2$	$K = 1 \cdot 10^{-31}$ cm⁶/s
	24	$e + O_2 \rightarrow O^- + O$	$k =$ $2 \cdot 10^{-10}$ cm³/s, $\varepsilon = 6.5\,eV$
Formation of free electrons	25	$O^- + O_3 \rightarrow e + 2O_2 + 2.6\,eV$	$k = 3 \cdot 10^{-10}$
	26	$O_2^- + O_2 \rightarrow e + O_3 + 0.62\,eV$	$k = 3 \cdot 10^{-10}$
	27	$O^- + O_2 \rightarrow e + O_3 - 0.42\,eV$	$k < 1 \cdot 10^{-12}$
Dissociation	28	$e + O_2 \rightarrow 2O + e$	$k = 2 \cdot 10^{-10}$, $\varepsilon > 6\,eV$

Table 6.6 Rates of collision processes involving of positive and negative ions in air. The rate constants k of pair processes are given in cm^3/s and correspond to room temperature

Type	Number	Process	Rate constant, cm^3/s
Charge exchange	1	$N_2^+ + O_2 \rightarrow N_2 + O_2^+ + 3.51\,eV$	$1 \cdot 10^{-10}$
	2	$O^+ + O_2 \rightarrow O + O_2^+ + 1.57\,eV$	$2 \cdot 10^{-11}$
	3	$N^+ + O_2 \rightarrow N + O_2^+ + 2.46\,eV$	$1.4 \cdot 10^{-10}$
	4	$NO + O_2^+ \rightarrow$ $NO^+ + O_2 + 2.81\,eV$	$4.4 \cdot 10^{-10}$
	5	$O^+ + NO \rightarrow O + NO^+ + 4.35\,eV$	$k < 1.3 \cdot 10^{-12}$
	6	$N_2^+ + H_2O \rightarrow$ $N_2 + H_2O^+ + 2.97\,eV$	$2.3 \cdot 10^{-9}\,s$
	7	$O_2^- + O \rightarrow O_2 + O^- + 1.02\,eV$	$3 \cdot 10^{-10}$
	8	$O^- + O_3 \rightarrow O + O_3^- + 0.65\,eV$	$8 \cdot 10^{-10}$
	9	$NO^- + O_2 \rightarrow$ $NO + O_2^- + 0.41\,eV$	$7 \cdot 10^{-10}$
	10	$O_2^- + O_3 \rightarrow O_2 + O_3^- + 1.67\,eV$	$6 \cdot 10^{-10}$
Ion-molecular reactions	11	$N_2^+ + O \rightarrow NO^+ + N + 3.05\,eV$	$1.4 \cdot 10^{-10}$
	12	$O^+ + N_2 \rightarrow NO^+ + N + 1.1\,eV$	$1.3 \cdot 10^{-12}$
	13	$N^+ + O_2 \rightarrow O^+ + NO + 2.3\,eV$	$4.5 \cdot 10^{-10}$
Charge exchange dissociation	14	$N_3^+ + O_2 \rightarrow N_2 + N + O_2^+ 0.5\,eV$	$7.4 \cdot 10^{-11}$
	15	$N_4^+ + O_2 \rightarrow 2N_2 + O_2^+ 2.6\,eV$	$3 \cdot 10^{-10}$
Exchange ion-molecular reactions	16	$N_4^+ + H_2O \rightarrow$ $2N_2 + H_2O^+ + 2.1\,eV$	$2.4 \cdot 10^{-9}$
	17	$O_2^+ \cdot N_2 + O_2 \rightarrow$ $O_4^+ + N_2 + 0.2\,eV$	$2.5 \cdot 10^{-11}$, $T = 80\,K$
	18	$O_2^+ \cdot N_2 + H_2O \rightarrow$ $O_2^+ \cdot H_2O + N_2 + 0.5\,eV$	$4 \cdot 10^{-9}$
	19	$O_4^+ + H_2O \rightarrow$ $O_2^+ \cdot H_2O + O_2 + 0.3\,eV$	$1.8 \cdot 10^{-9}$
	20	$H_3O^+ \cdot OH + H_2O \rightarrow$ $H_3O^+ \cdot H_2O + OH$	$2.3 \cdot 10^{-9}$
	21	$NH_4^+ \cdot H_2O + NH_3 \rightarrow$ $NH_4^+ \cdot NH_3 + H_2O$	$1.2 \cdot 10^{-9}$
	22	$O_3^- + NO_2 \rightarrow NO_3 + O^- + O_2$	$2.8 \cdot 10^{-10}$
	23	$O_4^- + O \rightarrow O_3^- + O_2$	$4 \cdot 10^{-10}$
	24	$O_2^- \cdot (H_2O)_2 + NO \rightarrow$ $NO_3^- \cdot H_2O + H_2O$	$2 \cdot 10^{-9}$

Table 6.7 Rates of collision processes in atmospheric plasma involving neutral atomic particles. The rate constants k of pair processes are given in cm^3/s and correspond to room temperature

Type	Number	Process	Time (τ) or rate constant (cm^3/s)
Radiation of metastable atoms	1	$O(^1S) \rightarrow O(^1D) + \hbar\omega$	$\tau = 0.8$ s, $\lambda = 558$ nm
	2	$O(^1D) \rightarrow O(^3P) + \hbar\omega$	$\tau = 140$ s, $\lambda = 630$ nm
	3	$N(^2D_{5/2}) \rightarrow N(^4S) + \hbar\omega$	$\tau = 1.4 \cdot 10^5$ s, $\lambda = 520$ nm
	4	$N(^2D_{3/2}) \rightarrow N(^4S) + \hbar\omega$	$\tau = 6 \cdot 10^4$ s, $\lambda = 520$ nm
Quenching of excited atoms and molecules	5	$2O_2(^1\Delta_g) \rightarrow O_2 + O_2(^1\Sigma_g^+)$	$k = 2 \cdot 10^{-17}$ cm^3/s
	7	$O_2(^1\Delta_g) + O_2 \rightarrow 2O_2$	$k = 2 \cdot 10^{-18}$ cm^3/s
	8	$O_2(^1\Sigma_g^+) + N_2 \rightarrow O_2 + N_2$	$k = 2 \cdot 10^{-17}$ cm^3/s
	9	$N_2(A^3\Sigma^+) + O_2 \rightarrow N_2 + O_2$	$k = 4 \cdot 10^{-12}$ cm^3/s
	10	$O(^1D) + O_2 \rightarrow O + O_2$	$k = 5 \cdot 10^{-11}$ cm^3/s
	11	$O(^1D) + N_2 \rightarrow O + N_2$	$k = 6 \cdot 10^{-11}$ cm^3/s
	12	$O(^1S) + O_2 \rightarrow O + O_2$	$k = 3 \cdot 10^{-13}$ cm^3/s
	13	$O(^1S) + O \rightarrow O + O$	$k = 7 \cdot 10^{-12}$ cm^3/s
Chemical reaction	14	$O + O_3 \rightarrow 2O_2$	$k = 7 \cdot 10^{-14}$ cm^3/s
Three body association	15	$O + 2O_2 \rightarrow O_3 + O_2$	$K = 7 \cdot 10^{-34}$ cm^6/s
	16	$O + O_2 + N_2 \rightarrow O_3 + N_2$	$K = 6 \cdot 10^{-34}$ cm^6/s

Table 6.8 The rate constants of three body processes involving positive and negative ions of air components. The rate constants are expressed in cm^6/s and correspond to room temperature if the temperature is not indicated

Number	Process	Rate constant, cm^6/s
1	$O_2^+ + 2O_2 \rightarrow O_4^+ + O_2$	10^{-29}, T = 200 K
2	$CO^+ + 2CO \rightarrow CO^+ \cdot CO + CO$	$1.3 \cdot 10^{-28}$
3	$CO_2^+ + 2CO_2 \rightarrow CO_2^+ \cdot CO_2 + CO_2$	$2.4 \cdot 10^{-28}$
4	$NO^+ + 2NO \rightarrow NO^+ \cdot NO + NO$	$5 \cdot 10^{-30}$
5	$CO_2^+ + 2CO \rightarrow CO_2^+ \cdot CO + CO$	$3.5 \cdot 10^{-28}$
6	$CO^+ \cdot CO + 2CO \rightarrow CO^+ \cdot (CO)_2 + CO$	$3.9 \cdot 10^{-30}$, T = 280 K
7	$O_2^+ + H_2O + N_2 \rightarrow O_2^+ \cdot H_2O + N_2$	$2.6 \cdot 10^{-28}$
8	$O_2^+ + H_2O + O_2 \rightarrow O_2^+ \cdot H_2O + O_2$	$2 \cdot 10^{-28}$
9	$NO_2^+ + H_2O + N_2 \rightarrow NO_2^+ \cdot H_2O + N_2$	$5 \cdot 10^{-28}$
10	$O^- + 2O_2 \rightarrow O_3^- + O_2$	$9 \cdot 10^{-31}$
11	$O_2^- + 2O_2 \rightarrow O_4^- + O_2$	$4 \cdot 10^{-31}$
12	$O^- + CO_2 + O_2 \rightarrow CO_3^- + O_2$	$2 \cdot 10^{-28}$
13	$O^- + 2CO_2 \rightarrow CO_3^- + CO_2$	$2 \cdot 10^{-29}$
14	$O^- + H_2O + O_2 \rightarrow O^- \cdot H_2O + O_2$	$1.2 \cdot 10^{-28}$
15	$O_2^- + H_2O + O_2 \rightarrow O_2^- \cdot H_2O + O_2$	$1.9 \cdot 10^{-28}$
16	$O_3^- + 2N_2 \rightarrow O_3^- \cdot N_2 + N_2$	$1.5 \cdot 10^{-31}$
17	$O_3^- + H_2O + O_2 \rightarrow O_3^- \cdot H_2O + O_2$	$2.4 \cdot 10^{-28}$
18	$CO_3^- + H_2O + O_2 \rightarrow CO_3^- \cdot H_2O + O_2$	$1 \cdot 10^{-28}$
19	$OH^- + CO_2 + O_2 \rightarrow HCO_3^- + O_2$	$7.6 \cdot 10^{-28}$
20	$OH^- + H_2O + O_2 \rightarrow OH^- \cdot H_2O + O_2$	$2.5 \cdot 10^{-28}$
21	$NO_2^- + H_2O + O_2 \rightarrow NO_2^- \cdot H_2O + O_2$	$1.6 \cdot 10^{-28}$
22	$NO_2^- + H_2O + NO \rightarrow NO_2^- \cdot H_2O + NO$	$1.5 \cdot 10^{-28}$
23	$O_2^- \cdot H_2O + H_2O + O_2 \rightarrow$ $O_2^- \cdot (H_2O)_2 + O_2$	$5 \cdot 10^{-28}$
24	$OH^- \cdot H_2O + H_2O + O_2 \rightarrow$ $OH^- \cdot (H_2O)_2 + O_2$	$3.5 \cdot 10^{-28}$
25	$O^- + O_2^+ + N_2 \rightarrow O + O_2 + N_2$	$7 \cdot 10^{-26}$
26	$NO_2^- + NO^+ + N_2 \rightarrow NO + NO_2 + N_2$	$1 \cdot 10^{-25}$

Chapter 7
Transport Phenomena in Gaseous Systems

Abstract Values of transport coefficients of gases are given and include diffusion of atoms, thermal conductivity and viscosity of gases. Parameters of electron and ion drift in gases in an external electric field are represented.

7.1 Transport Coefficients of Gases

Transport coefficients characterize the connection between weak gradients of some quantities and fluxes. The diffusion coefficient D for atoms or molecules in a gas is the proportionality coefficient between the flux $\mathbf{j}$ of these atoms and the gradient of their concentration c, that is

$$\mathbf{j} = -DN_a\nabla c, \tag{7.1}$$

where N_a is the total number density of gas atoms or molecules. If the concentration of atoms of a given type is small ($c_i \ll 1$), i.e. this component is a small admixture to a gas, this formula may be reduced to the form

$$\mathbf{j} = -D_i\nabla N_i, \tag{7.2}$$

where N_i is the number density of atoms of a given component. The thermal conductivity coefficient of a gas κ is defined as the proportionality coefficient between the thermal flux $\mathbf{q}$ and the temperature gradient ∇T, i.e.

$$\mathbf{q} = -\kappa\nabla T \tag{7.3}$$

The viscosity coefficient η is the proportionality coefficient between the friction force F per square unit of a moving gas and the gradient of the average gas velocity. In the frame of reference where the direction of the average gas velocity $\mathbf{w}$ is x and the average velocity varies in the direction z, the friction force is proportional to the quantity $\partial w_x/\partial z$ and acts on the surface xy. Correspondingly, the viscosity coefficient η is defined as

$$F = -\eta\frac{\partial w_x}{\partial z}, \tag{7.4}$$

and this definition holds true both for a gas and for a liquid.

© Springer International Publishing AG, part of Springer Nature 2018
B. M. Smirnov, *Atomic Particles and Atom Systems*, Springer Series on Atomic,
Optical, and Plasma Physics 51, https://doi.org/10.1007/978-3-319-75405-5_7

Transport coefficients in a gas are determined by collision processes. Because the elastic cross section of collision of atomic particles in an atomic or molecular gas at not high temperatures is large compared to cross sections of inelastic processes, the transport coefficients of a gas are expressed through average cross sections of elastic collisions of atomic particles. The simple connection between these quantities takes place in the Chapman-Enskog approximation [154–156] that corresponds to an expansion of corresponding values over a small numerical parameter, and the accuracy of the first Chapman-Enskog approximation is estimated as several percent. The diffusion coefficient D of a test atom or molecule in a gas is given in the first Chapman-Enskog approximation by [154, 155]

$$D = \frac{3\sqrt{\pi T}}{8N\sqrt{2\mu\bar{\sigma}}}, \quad \bar{\sigma} \equiv \Omega^{(1,1)}(T) = \frac{1}{2}\int\limits_0^\infty e^{-t}t^2\sigma^*(t)\mathrm{d}t, \quad t = \frac{\mu g^2}{2T} \qquad (7.5)$$

Here T is the gas temperature expressed in energetic units, N is the number density of gas atoms, μ is the reduced mass of a test atom and a gas atom, $\sigma^*(g)$ is the diffusion cross section of collision of these atomic particles at a relative velocity g of collision; brackets mean an average over atom velocities with the Maxwell distribution function.

The thermal conductivity coefficient κ in the first Chapman-Enskog approximation is given by

$$\kappa = \frac{25\sqrt{\pi T}}{32\bar{\sigma_2}\sqrt{m}} \qquad (7.6)$$

where m is the mass of an atom or molecule, an average of the cross section of elastic collision is made on the basis of the following formula

$$\bar{\sigma_2} \equiv \Omega^{(2,2)}(T) = \int\limits_0^\infty t^2\exp(-t)\sigma^{(2)}(t)\mathrm{d}t, \quad t = \frac{\mu g^2}{2T}, \quad \sigma^{(2)}(T) = \int(1 - \cos^2\vartheta)\mathrm{d}\sigma,$$
$$(7.7)$$

The viscosity coefficient η of the first Chapman-Enskog approximation is determined by formula

$$\eta = \frac{5\sqrt{\pi T m}}{24\bar{\sigma_2}}, \qquad (7.8)$$

where the average cross section $\bar{\sigma_2}$ is given by formula (7.7). In this approximation the coefficients of thermal conductivity and viscosity are connected by the relation

$$\kappa = \frac{15}{4m}\eta \qquad (7.9)$$

Formulas (7.5), (7.6), (7.8) for the first Chapman-Enskog approximation are simplified if particle collision is described by the hard sphere model (6.3). Then the average cross sections are given by

$$\bar{\sigma} = \sigma_o; \ \overline{\sigma_2} = \frac{2}{3}\sigma_o, \tag{7.10}$$

where $\sigma_o = \pi R_o^2$, and R_o is the hard sphere radius. This leads to the following expressions for the transport coefficients instead of formulas (7.5), (7.6), (7.8)

$$D = \frac{3\sqrt{\pi T}}{8\sqrt{2\mu}N_a\sigma_o}; \ \kappa = \frac{75\sqrt{\pi T}}{64\sigma_o\sqrt{m}} \ \eta = \frac{15\sqrt{\pi T m}}{48\sigma_o}, \tag{7.11}$$

Values of the coefficients of self-diffusion are given in Table 7.1 and are reduced as usually to the normal density of atoms or molecules $N_a = 2.687 \cdot 10^{19}$ cm^{-3} ($T = 273$ K, $p = 1$ atm) because the diffusion coefficient is inversely proportional to the number density N_a of gas atoms. Approximating the temperature dependence of the diffusion coefficient as

$$D = D_o \left(\frac{T}{300}\right)^{\gamma}, \tag{7.12}$$

where the temperature is expressed in K, and the parameters of this formula are based on the data [157, 158] for the temperature range T = 300–1500 K and are given in Table 7.2. Note that within the framework of the hard sphere model according to formula (7.11) we have $\gamma = 1.5$ since the number density of atoms is inversely proportional to the temperature. In addition to these data, the diffusion coefficients of metal atoms in inert gases at temperature $T = 500$ K are given in Table 7.3.

The gas-kinetic cross section is introduced within the framework of the hard sphere model on the basis of the expression for the diffusion coefficient (7.11), whose value is taken at room temperature. We have the following relation between the gas-kinetic cross section σ_g and the diffusion coefficient D

$$\sigma_g = \frac{0.47}{N_a D}\sqrt{\frac{T}{\mu}} \tag{7.13}$$

where N_a is the number density of atoms or molecules, D is the diffusion coefficient, T is room temperature expressed in energetic units, μ is the reduced mass of colliding particles. The values of gas-kinetic cross sections σ_g defined on the basis of formula (7.13) are given in Table 7.4.

Values of the thermal conductivity coefficients of inert gases are given in Table 7.5 at pressure of 1 atm, and Table 7.6 contains the viscosity coefficients for these gases at atmospheric pressure where the pressure dependence is weak.

In addition, these transport coefficients in the first Chapman-Enskog approximation [154, 155] are connected by the relation

$$\kappa = \frac{5c_V\eta}{2m} \tag{7.14}$$

Here κ is the thermal conductivity coefficient, η is the viscosity coefficient, c_V is the heat capacity per atom or molecule that for ideal gas is $c_V = 3/2$, and m is the mass of an atom or molecule.

Table 7.7 lists the diffusion coefficients of excited atoms and molecules in a parent gas. Comparison with data of Table 7.1 exhibits that diffusion coefficients of excited atoms exceeds those for atoms in the ground state because of a more strong interaction.

Table 7.8 gives the values of conversational factors for units in which transport coefficients in gases are represented.

Explanations to Table 7.8.

1. The coefficient of thermal conductivity of gases in the first Chapman-Enskog approximation is $\kappa = 25\sqrt{\pi T}/(32\sqrt{m}\,\overline{\sigma_2})$. Here T is the gas temperature, m is the atom or molecule mass, $\overline{\sigma_2}$ is the average cross section of collision between gas atoms or molecules in acordance with formula (7.7).
2. The gas viscosity in the first Chapman-Enskog approximation $\eta = 5\sqrt{\pi T m}/(24\overline{\sigma_2})$; notations are the same as above.

7.2 Ion Drift in Gas in External Electric Field

In consideration the ion motion in a gas under the action of an external electric field of strength E, we restrict ourselves by limiting cases of low and high electric fields. The limit of low electric fields corresponds to the criterion

$$eE\lambda \ll T, \tag{7.15}$$

where $\lambda = 1/(N_a\sigma)$ is the mean free path of ions in a gas, σ is the cross section of ion-atom scattering on large angles, N_a is the number density of atoms, and we assume the ion and atom masses to have the same order of magnitude. In this limiting case the ion drift velocity, i.e. its average velocity, is proportional to the electric field strength

$$w = EK, \tag{7.16}$$

and this is the definition of the ion mobility K at low electric fields where it is independent of the electric field strength.

The diffusion coefficient of a charged particle in a gas D is connected with the ion mobility K by the Einstein relation [161, 162]

$$D = \frac{KT}{e} \tag{7.17}$$

Here T is the gas temperature, and the Einstein relation follows from the equilibrium condition that leads to equality of the diffusion and drift fluxes. Note that though this relation is called the Einstein relation, it was derived by Townsend several years before [163], and Einstein used these results in the analysis of Brownian motion of particles [164]. The Einstein relation (7.17) allows us to find the following formula for ion mobility K on the basis of expression (7.5) for the diffusion coefficient

$$K = \frac{3\sqrt{\pi e}}{8\sqrt{2\mu}N_a\sigma^*}, \tag{7.18}$$

where σ^* is the diffusion cross section for ion-atom collision.

The diffusion cross section of ion-atom scattering under the action of the polarization interaction potential (5.5) between them is close to the capture cross section and is equal to [134, 165]

$$\sigma^*(v) = \int (1 - \cos\vartheta)d\sigma = 2.21\pi\sqrt{\frac{\alpha e^2}{\mu g^2}}, \tag{7.19}$$

i.e. exceeds the capture cross section by about 10%. For this cross section the ion mobility reduced to the number density of gas atoms $N_a = 2.69 \cdot 10^{19}$ cm^{-3} is given by the Dalgarno formula [116, 165]

$$K = \frac{36}{\sqrt{\alpha\mu}} \frac{\text{cm}^2}{\text{V s}}, \tag{7.20}$$

where the ion-atom reduced mass is expressed in atomic mass units ($1.66 \cdot 10^{-24}$ g), and the polarizability is given in atomic units (a_o^3). It is of importance that this mobility does not depend on the temperature, and ion parameters are taken into account only by the ion-atom reduced mass.

Table 7.9 contains the mobilities of alkali metal atoms in inert gases and nitrogen at room temperature in zero electric field [166–171]. These data allow us to check the accuracy of the Dalgarno formula (7.20) for the mobility of ions in a foreign gas. Indeed, Table 7.10 contains the ratio of experimental mobilities of alkali ions in inert gases and nitrogen at room temperature [166–171] to those calculated by formula (7.20). One can see that experimental data exceeds the theoretical ones, and the statistical treatment gives for the average value of this ratio 1.15 ± 0.10. From this one can obtain that the Dalgarno formula (7.20) may be used for a rough evaluation of the ion mobility of ions. One can correct the Dalgarno formula with taking into account experimental data, and then we have

$$K = \frac{41 \pm 4}{\sqrt{\alpha\mu}} \frac{\text{cm}^2}{\text{V s}}, \tag{7.21}$$

and the indicated accuracy ($\sim$10%) is the statistical one, and the real divergence from formula (7.21) may be more. In particular, he ratio of the mobility of alkali metal ions in helium to that given by formula (7.21) is equal to 1.13 ± 0.06. Thus, one can suggest formula (7.21) for a rough evaluation of the ion mobility in gases and conclude from this that the ion-atom polarization interaction gives the basic contribution to the ion mobility in a gas.

On the basis of the Einstein relation and formula (7.21) we have for the ion diffusion coefficient in a gas

$$D = \frac{1.0 \pm 0.1}{\sqrt{\alpha\mu}} \frac{\text{cm}^2}{\text{s}}, \qquad (7.22)$$

which is taken as usually under normal conditions.

If atomic ions are moving in a parent atomic gas in an electric field, scattering of ions has a specific character according to the Sena effect [125, 126], as it is shown in Fig. 7.1. Then colliding ion and atom move along straightforward trajectories, but the process of charge exchange changes a charged particle, so that electron transfer to another atomic rest leads to effective ion scattering. This takes place, in particular, at room gas temperature, where the elastic cross section of ion-atom scattering is less than the cross section of resonant charge exchange. Then the cross section of resonant charge exchange determines the mobility of ions in a parent gas.

Indeed, in absence of elastic ion-atom scattering in the center-of-mass frame of reference the ion scattering angle is $\vartheta = \pi$, and the diffusion cross section of ion scattering in absence of elastic scattering is [172]

$$\sigma^* = \int (1 - \cos\vartheta)\,d\sigma = 2\sigma_{res}, \qquad (7.23)$$

where σ_{res} is the cross section of the resonant charge exchange process. Then, assuming the cross section σ_{res} of resonant charge exchange to be independent of the collision velocity, we obtain from formula (7.18) for the ion mobility in a parent gas in the first Chapman-Enskog approach [173]

$$K_I = \frac{0.331e}{N_a\sqrt{Tm_a}\sigma_{res}(2.2v_T)} \qquad (7.24)$$

where the collision velocity is $v_T = \sqrt{2T/m_a}$, and the argument of the cross section of resonant charge exchange σ_{res} indicates a velocity at which this cross section is taken. The second Chapman-Enskog approach [174] exceeds the ion mobility by 1/40 and gives for the ion mobility in a parent gas in the absence of elastic scattering [134, 151]

$$K = \frac{0.341e}{N_a\sqrt{Tm_a}\sigma_{res}(2.1v_T)} \qquad (7.25)$$

Reducing the mobility to the normal number density of atoms $N_a = 2.69 \cdot 10^{19}$ cm^{-3}, one can rewrite formula (7.25) to the form [134, 151]

$$K = \frac{1340}{\sqrt{Tm_a}\sigma_{res}(2.1v_T)} \frac{cm^2}{Vs} \tag{7.26}$$

where the temperature T is given in Kelvin, the atom and ion mass $m_a = M$ are equal and are expressed in units of atomic masses, and the resonant charge exchange cross section is given in 10^{-15} cm^2; the argument indicates the collision velocity that corresponds to the collision energy $4.5T$ in the laboratory frame of reference. Diagram of Fig. 7.2 contains the values of the mobilities of atomic ions in parent gases at temperatures $T = 300, 800$ K. The accuracy of this formula is approximately 10%.

Accounting for elastic ion-atom scattering due to the polarization ion-atom interaction, we obtain for the ion mobility in a parent gas at zero electric field [132, 134]

$$K = \frac{K_o}{1 + x + x^2}, \quad x = \frac{\sqrt{\alpha e^2 / T}}{\sigma_{res}(\sqrt{7.5T/M})} \tag{7.27}$$

where K_o is the mobility in ignoring elastic scattering according to formula (7.25), α is the atom polarizability. This formula accounts for transition to the limits of low and high temperatures. The contribution to the ion mobility in a parent gas of elastic ion-atom scattering due to the polarization interaction is given in Table 7.11 for inert gases and alkali metal vapors [134]. Note that taking into account elastic scattering leads to a decrease the ion mobility.

The diffusion coefficient of ions in a parent gas at zero field follows from the Einstein relation (7.17) and formula (7.24) is given by

$$D = 0.12\sqrt{\frac{T}{M}} \frac{1}{\sigma_{res}(\sqrt{9T/M})}, \tag{7.28}$$

where the diffusion coefficient is expressed in cm^2/s, and other are explained in formula (7.24).

In the case of strong electric fields

$$eE\lambda \gg T, \tag{7.29}$$

the velocity distribution function of ions in the electric field direction x has the following form, if we ignore elastic ion-atom scattering and assume the cross section of resonant charge exchange to be independent of the collision velocity [125]

$$f(v_x) = C \exp\left(-\frac{Mv_x^2}{2eE\lambda}\right), \quad v_x > 0 \tag{7.30}$$

Here C is the normalization constant, and this formula gives for the ion drift velocity w_i and the average ion energy

$$w_i = \overline{v_x} = \sqrt{\frac{2eE\lambda}{\pi M}}, \quad \overline{\varepsilon} = \frac{M\overline{v_x^2}}{2} = \frac{eE\lambda}{2} \tag{7.31}$$

Accounting for the velocity dependence for the cross section of resonant charge exchange, it is taken in formula (7.31) at the ion velocity $1.4\sqrt{eE\lambda/M}$ in the laboratory frame of reference (an atom is motionless). The diffusion ion coefficients in the field direction $D_\parallel$ and in the perpendicular direction to it $D_\perp$ are given by [35]

$$D_\parallel = 0.137\lambda w, \ \ D_\perp = \frac{T\lambda}{Mw}, \ \frac{D_\parallel}{D_\perp} = 0.137\frac{Mw^2}{T} = 0.087\frac{eE\lambda}{T} \tag{7.32}$$

The ion drift velocity is given by formulas (7.15) and (7.29) in the limit of low electric field strengths and by formula (7.31) in the limit of largeelectric field strengths. One can construct the ion drift velocity in an intermediate range of electric fields. Indeed, let us introduce the parameter

$$\beta = \frac{eE}{TN_a\sigma_{res}}, \tag{7.33}$$

and in the first approximation assume the cross section of resonant charge exchange to be independent of the collision velocity. Let us represent the ion drift velocity in the form

$$w = \sqrt{\frac{2T}{m}}\Phi(\beta), \tag{7.34}$$

and according to formulas (7.24) and (7.31) the limiting dependencies for the function $\Phi(\beta)$ have the form

$$\Phi(\beta) = 0.48\beta, \ \beta \ll 1; \ \ \Phi(\beta) = \sqrt{\frac{\beta}{\pi}}, \ \beta \gg 1 \tag{7.35}$$

As follows from the solution of the kinetic equation [175, 176] for the ion distribution function on velocities, the reduced ion drift velocity $\Phi(\beta)$ may be approximated by the dependence [134, 176]

$$\Phi(\beta) = \frac{0.48\beta}{(1 + 0.22\beta^{3/2})^{1/3}} \tag{7.36}$$

The dependence $\Phi(\beta)$ for intermediate values β is given in Fig. 7.3.
    Table 7.12 contains conversational factors between transport coefficients. Explanation to Table 7.12.

1. The Einstein relation (7.17) for the diffusion coefficient of a charged particle in a gas $D = KT/e$, where D, K are the diffusion coefficient and mobility of a charged particle, T is the gas temperature.

2. The Einstein relation (7.17) for the mobility of a charged particle in a gas $K = eD/T$.

3. The diffusion coefficient of an atomic particle in a gas in the first Chapman-Enskog approximation according to formula (7.5) $D = 3\sqrt{2\pi T/\mu}/(16N_a\overline{\sigma_1})$, where T is the gas temperature, N_a is the number density of gas atoms or molecules, μ is the reduced mass of a colliding particle and gas atom or molecule, $\overline{\sigma_1}$ is the average cross section of collision.

4. The mobility of a charged particle in a gas in the first Chapman-Enskog approximation according to formula (7.18) $K = 3e\sqrt{2\pi/(T\mu)}/(16N_a\overline{\sigma_1})$; notations are the same as above.

5. The parameter of ion drift in a gas in a constant electric field $\xi = eE/(TN\sigma)$, where E is the electric field strength, T is the gas temperature, N is the number density of atoms or molecules, σ is the cross section of collision.

7.3 Electron Transport in Gases

If an electron is located in a gas in external electric field, its drift is restricted by electron-atom elastic collisions. The cross section of electron-atom scattering at energies up to several eV is determined by a finite number of collision momenta, and hence this process has a quantum character. Table 7.13 contains reduced zero-field mobilities of electrons in inert gases at room temperature $K_e N_a$ (K_e is the electron mobility, N_a is the number density of gas atoms) and the values of the reduced diffusion coefficients of electrons $D_e N_a$ (D_e is the diffusion coefficient of electrons) which are connected with the electron mobilities by the Einstein relation (7.17).

Being moved in an atomic gas in an external electric field, electrons obtain energy from the field and transfer it to atoms in elastic collisions until the electron energy does not allow to excite atoms. In each strong collision an electron is scattered on a large angle, so that the electron momentum changes significantly, whereas its energy varies weakly. Therefore the electron distribution function on velocities is close to a spherically symmetric one. Hence, the electron drift velocity w and the diffusion coefficient for an electron in a gas $D_\perp$ in the perpendicular direction to the field are expressed through the spherical distribution function of electrons on the basis of following expressions [177]

$$w = \frac{eE}{3m_e}\left\langle \frac{1}{v^2}\frac{d}{dv}\left(\frac{v^3}{\nu}\right)\right\rangle, \quad D_\perp = \frac{1}{3}\left\langle\left(\frac{v^2}{\nu}\right)\right\rangle \tag{7.37}$$

Here brackets mean an average with the spherical distribution function on electron velocities, and the rate of electron elastic scattering in a gas is $\nu = N_a v\sigma^*(v)$, where N_a is the number density of atoms, v is the electron velocity. In the simplest case where the rate of electron elastic scattering ν is independent of the electron velocity v, these formulas take the form

$$w_e = \frac{eE}{m_e \nu}, \quad D_e = \frac{\langle v^2 \rangle}{\nu}, \qquad (7.38)$$

The data of Table 7.14 for the electron drift velocity in some atomic and molecular gases in an external electric field show that this case is not realistic. Note that the drift velocity depends on the reduced electric field strength E/N_a, where E is the electric field strength, and the unit for this quantity is *Townsend* 1 Td = 10^{-17} V cm^2.

This character of electron processes when an electron takes an energy from an external electric field and then transfers it to atoms as a result of elastic and inelastic collisions takes place in a gas discharge plasma. A gas discharge plasma results from passage of electric current through a gas under the action of an external electric field. Collision of electrons with atoms in a gas discharge plasma establish a self-maintaining equilibrium inside it.

Electron-atom collisions in gases and plasmas are of importance both for plasma processes and for transport processes in ionized gases [111, 181, 182]. The basic process among these ones is drift of electrons in gases under the action of an external electric field. Figures 7.4, 7.5, 7.6, and 7.7 give the dependence of the electron drift velocity w_e on the specific electric field strength E/N_a (E is the strength of an external electric field, N_a is the number density of gas atoms or molecules) in helium, argon, xenon and mercury correspondingly. Note that according to these experimental data in all the cases one can extract three ranges of the specific electric field strength. In the range 1 an external field is relatively small and the average energies of electrons and atoms are identical. In the second range, the saturation range, the average electron energy is small compared to the atom excitation energy, and the electron drift velocity results from elastic electron-atom collisions. In the range 3 inelastic electron-atom collisions become essential, and the power resulted from electron interaction with an external electric field is consumed on atom excitation.

One can introduce the Townsend characteristic energy [185] which is the ratio of the electron diffusion coefficient to the its mobility K_e that is defined as $K_e = w_e/E$. It is convenient to define the Townsend characteristic energy as

$$T_{ef} = \frac{eD_\perp}{K_e} = \frac{eED_\perp}{w_e}, \qquad (7.39)$$

since in the limit of low electric field strengths according to the Einstein relation the ratio $eD_\perp/K_e$ is the gas temperature expressed in energetic units. At arbitrary electric field strengths the value eD_e/K_e characterizes a typical electron energy. Figure 7.8 gives the dependence of this value on the electric field strength in the case of electron drift in argon.

In addition, Tables 7.15 and 7.16 contains transport coefficients of electrons in in helium and argon under the action of an electric field [187]. The number density is small, and transport of electrons in these gases is determined by collisions of electrons with atoms. The limit of low electric field strength where electrons are characterized by the Maxwell velocity distribution function with the gaseous temperature, corresponds to the criterion $E/N_a \ll 0.003$ Td in the helium case and at $E/N_a \ll 3 \cdot 10^{-4}$ Td in the argon case.

Figures

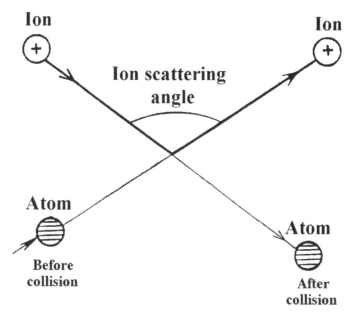

Fig. 7.1 Atom and ion are moving along straightforward trajectories, and resonant charge exchange takes place, so that a valence electron transfers to another atomic rest. This leads to effective ion scattering within the framework of the Sena effect

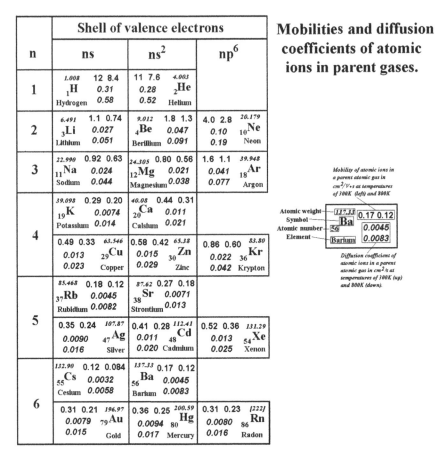

Mobilities and diffusion coefficients of atomic ions in parent gases.

Fig. 7.2   Mobilities of atomic ions in parent gases

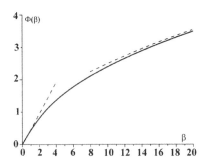

Fig. 7.3 The dependence of the reduced drift velocity of atomic ions in a parent gas Φ on the reduced electric field strength β (solid curve) in accordance with formula (7.36). Dotted curves correspond to limiting cases

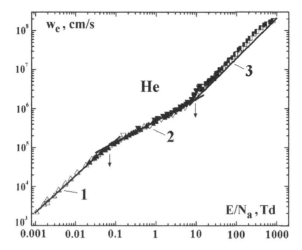

Fig. 7.4 Electron drift velocity w_e in helium as a function of the specific electric field strength

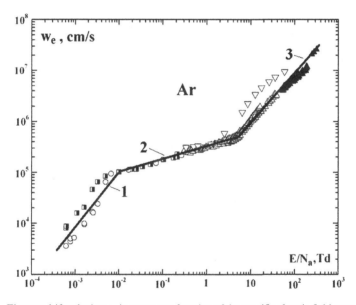

Fig. 7.5 Electron drift velocity w_e in argon as a function of the specific electric field strength [178]

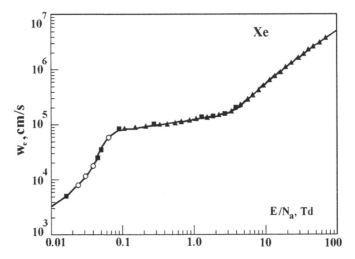

Fig. 7.6 Electron drift velocity w_e in xenon as a function of the specific electric field strength. Circles [179], triangles and squares [180], solid curve [147]

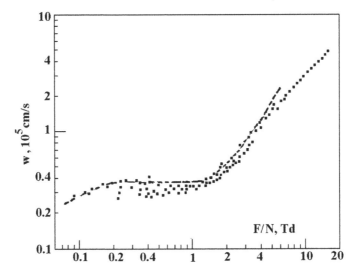

Fig. 7.7 Electron drift velocity w_e in a mercury vapor as a function of the specific electric field strength according to measurements [183, 184]

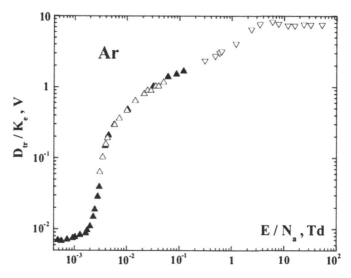

Fig. 7.8 Townsend characteristic energy D_e/K_e in argon as the function of the specific electric field strength [178]

Tables

Table 7.1 The diffusion coefficients D of atoms and molecules in a parent gas given in cm^2/s and reduced to the normal number density

Gas	D, cm^2/s	Gas	D, cm^2/s	Gas	D, cm^2/s
He	1.6	H_2	1.3	H_2O	0.28
Ne	0.45	N_2	0.18	CO_2	0.096
Ar	0.16	O_2	0.18	NH_3	0.25
Kr	0.084	CO	0.18	CH_4	0.20
Xe	0.048				

Table 7.2 The parameters of formula (7.12) with the diffusion coefficient D_o at temperature $T = 300$ K is expressed in cm^2/s, and values of the parameter γ of formula (7.12) are given in parentheses

Atom, gas	He	Ne	Ar	Kr	Xe
He	1.66(1.73)	1.08(1.72)	0.76(1.71)	0.68(1.67)	0.56(1.66)
Ne		0.52(1.69)	0.32(1.72)	0.25(1.74)	0.23(1.67)
Ar			0.195(1.70)	0.15(1.70)	0.115(1.74)
Kr				0.099(1.78)	0.080(1.79)
Xe					0.059(1.76)
Average					(1.72 ± 0.03)

Table 7.3 The diffusion coefficient of metal atoms in inert gases at temperature 300 K, expressed in cm^2/s [159]

Atom, gas	He	Ne	Ar	Kr	Xe
Li	0.54	0.29	0.28	0.24	0.20
Na	0.50	0.31	0.18	0.14	0.12
K	0.54	0.24	0.22	0.11	0.087
Cu	0.68	0.25	0.11	0.077	0.059
Rb	0.39	0.21	0.14	0.11	0.088
Cs	0.34	0.22	0.13	0.080	0.057
Hg	0.46	0.21	0.11	0.072	0.056

Table 7.4 Gas-kinetic cross sections for collisions of atoms or molecules σ_g expressed in 10^{-15} cm^2

Colliding particles	He	Ne	Ar	Kr	Xe	H$_2$	N$_2$	O$_2$	CO	CO$_2$
He	1.3	1.5	2.1	2.4	2.7	1.7	2.3	2.2	2.2	2.7
Ne		1.8	2.6	3.0	3.3	2.0	2.4	2.5	2.7	3.1
Ar			3.7	4.2	5.0	2.7	3.8	3.9	4.0	4.4
Kr				4.8	5.7	3.2	4.4	4.3	4.4	4.9
Xe					6.8	3.7	5.0	5.2	5.0	6.1
H$_2$						2.0	2.8	2.7	2.9	3.4
N$_2$							3.9	3.8	3.7	4.9
O$_2$								3.8	3.7	4.4
CO									4.0	4.9
CO$_2$										7.5

Table 7.5 Thermal conductivity coefficients in gases κ expressed in 10^{-4} W/(cm K) [160]

T, K	100	200	300	400	600	800	1000
H$_2$	6.7	13.1	18.3	22.6	30.5	37.8	44.8
He	7.2	11.5	15.1	18.4	25.0	30.4	35.4
CH$_4$	–	2.17	3.41	4.88	8.22	–	–
NH$_3$	–	1.53	2.47	6.70	6.70	–	–
H$_2$O	–	–	–	2.63	4.59	7.03	9.74
Ne	2.23	3.67	4.89	6.01	7.97	9.71	11.3
CO	0.84	1.72	2.49	3.16	4.40	5.54	6.61
N$_2$	0.96	1.83	2.59	3.27	4.46	5.48	6.47
Air	0.95	1.83	2.62	3.28	4.69	5.73	6.67
O$_2$	0.92	1.83	2.66	3.30	4.73	5.89	7.10
Ar	0.66	1.26	1.77	2.22	3.07	3.74	4.36
CO$_2$	–	0.94	1.66	2.43	4.07	5.51	6.82
Kr	–	0.65	1.00	1.26	1.75	2.21	2.62
Xe	–	0.39	0.58	0.74	1.05	1.35	1.64

Table 7.6 The viscosity coefficient of gases at atmospheric pressure given in 10^{-5} g/(cm s) [157]

T, K	100	200	300	400	600	800	1000
H_2	4.21	6.81	8.96	10.8	14.2	17.3	20.1
He	9.77	15.4	19.6	23.8	31.4	38.2	44.5
CH_4	–	7.75	11.1	14.1	19.3	–	–
H_2O	–	–	–	13.2	21.4	29.5	37.6
Ne	14.8	24.1	31.8	38.8	50.6	60.8	70.2
CO	–	12.7	17.7	21.8	28.6	34.3	39.2
N_2	6.88	12.9	17.8	22.0	29.1	34.9	40.0
Air	7.11	13.2	18.5	23.0	30.6	37.0	42.4
O_2	7.64	14.8	20.7	25.8	34.4	41.5	47.7
Ar	8.30	16.0	22.7	28.9	38.9	47.4	55.1
CO_2	–	9.4	14.9	19.4	27.3	33.8	39.5
Kr	–	–	25.6	33.1	45.7	54.7	64.6
Xe	–	–	23.3	30.8	43.6	54.7	64.6

Table 7.7 Diffusion coefficients of metastable atoms or molecules in parent gases at room temperature [135]

Metastable atom, molecule	D, cm^2/s
$He(2^3S)$	0.59
$He(2^1S)$	0.52
$He(2^3\Sigma_u)$	0.45
$Ne(^3P_2)$	0.20
$Ne(^3P_0)$	0.20
$Ar(^3P_2)$	0.067
$Ar(^3P_0)$	0.071
$Kr(^3P_2)$	0.039
$Kr(^3P_0)$	0.043
$Xe(^3P_2)$	0.024
$Xe(^3P_0)$	0.020
$N(^2D)$	0.24
$N(^2D)$	0.19
$N_2(A^3\Sigma_u)$	0.20
$O_2(^1\Delta_g)$	0.19

Table 7.8 Conversion factors for transport coefficients in gases

Number	Formula	Factor C	Units
1.	$\kappa = C\sqrt{T/m}/\overline{\sigma_2}$	$1.743 \cdot 10^4$ W/(cm K)	T in K, m in a.u.m., $\overline{\sigma_2}$ in Å^2
		$7.443 \cdot 10^5$ W/(cm K)	T in K, m in e.u.m., $\overline{\sigma_2}$ in Å^2
2.	$\eta = C\sqrt{Tm}/\overline{\sigma_2}$	$5.591 \cdot 10^{-5}$ g/(cm s)	T in K, m in a.u.m., $\overline{\sigma_2}$ in Å^2

Table 7.9 The zero field mobility (in $cm^2/(V\ s)$) of positive ions of alkali metals in inert gases and nitrogen at room temperature [166, 167, 171]

Ion, gas	He	Ne	Ar	Kr	Xe	H_2	N_2
Li^+	22.9	10.6	4.6	3.7	2.8	12.4	4.15
Na^+	22.8	8.2	3.1	2.2	1.7	12.2	2.85
K^+	21.5	7.4	2.7	1.83	1.35	13.1	2.53
Rb^+	20	6.5	2.3	1.45	1.0	13.0	2.28
Cs^+	18.3	6.0	2.1	1.3	0.91	12.9	2.2

Table 7.10 The ratio between experimental and theoretical mobilities of ions in inert gases [134]

Ion, gas	He	Ne	Ar	Kr	Xe	H_2	N_2
Li^+	1.27	1.17	1.06	1.07	1.06	1.02	0.89
Na^+	1.38	1.24	1.07	1.06	1.09	1.13	0.97
K^+	1.35	1.28	1.08	1.10	1.07	1.14	0.98
Rb^+	1.29	1.26	1.08	1.07	1.0	1.14	0.99
Cs^+	1.19	1.18	1.07	1.08	1.07	1.15	0.99

Table 7.11 The contribution of elastic ion-atom scattering to the mobility of atomic ions in a parent gas at room temperature. K is the mobility without elastic scattering, ΔK is a decrease of the mobility due to elastic ion-atom scattering [134]

Ion, gas	$\Delta K/K$, %	Ion, gas	$\Delta K/K$, %
He	5.8	Li	12
Ne	7.8	Na	8.2
Ar	11	K	8.9
Kr	6.0	Rb	7.7
Xe	9.2	Cs	7.5

Table 7.12 Conversion factors for transport coefficients

Number	Formula	Factor C	Units
1.	$D = CKT$	$8.617 \cdot 10^{-5}$ cm^2/s	K in cm^2/(V s), T in K
		1 cm^2/s	K in cm^2/(V s), T in eV
2.	$K = CD/T$	11604 cm^2/(V s)	D in cm^2/s, T in K
		1 cm^2/(V s),	D in cm^2/s, T in eV
3.	$D = C\sqrt{T/\mu}/(N\overline{\sigma_1})$	$4.285 \cdot 10^{19}$ cm^2/s	$\overline{\sigma_1}$ in Å^2, N in cm^{-3}, T in K, μ in a.u.m.
		1.595 cm^2/s	$\overline{\sigma_1}$ Å^2, $N = 2.687 \cdot 10^{19}$ cm^{-3}, T in K, μ in a.u.m.
		171.8 cm^2/s	$\overline{\sigma_1}$ in Å^2, $N = 2.687 \cdot 10^{19}$ cm^{-3}, T in eV, μ in a.u.m.
		4.024 cm^2/s	$\overline{\sigma_1}$ in Å^2, $N = 2.687 \cdot 10^{19}$ cm^{-3}, T in eV, μ in e.u.m.
		7335 cm^2/s	$\overline{\sigma_1}$ in Å^2, $N = 2.687 \cdot 10^{19}$ cm^{-3}, T in eV, μ in e.u.m.
4.	$K = C(\sqrt{T\mu}N\overline{\sigma_1})^{-1}$	171.8 cm^2/(V s)	$\overline{\sigma_1}$ in Å^2, $N = 2.687 \cdot 10^{19}$ cm^{-3}, T in eV, μ in a.u.m.
		7338 cm^2/(V s)	$\overline{\sigma_1}$ in Å^2, $N = 2.687 \cdot 10^{19}$ cm^{-3}, T in eV, μ in e.u.m.
5.	$\xi = CE/(TN\sigma)$	$1.160 \cdot 10^{20}$	E in V/cm, T in K, σ in Å^2, N in cm^{-3}
		$1 \cdot 10^{16}$	E in V/cm, T in eV, σ in Å^2, N in cm^{-3}

Table 7.13 Parameters of electron drift in inert gas atoms at zero electric field and room temperature [178]

Gas	He	Ne	Ar	Kr	Xe
$D_e N_a$, 10^{21} cm^{-1} s^{-1}	7.2	72	30	1.6	0.43
$K_e N_a$, 10^{23} (cm s V)$^{-1}$	2.9	29	12	0.62	0.17

Table 7.14 The electron drift velocity w in gases expressed in 10^5 cm/s [178]

E/N, Td	He	Ne	Ar	Kr	Xe	H_2	N_2
0.001	0.028	0.35	0.15	0.015	0.004	0.020	–
0.003	0.080	0.55	0.53	0.038	0.013	0.052	–
0.01	0.25	0.89	1.0	0.13	0.040	0.17	0.39
0.03	0.64	1.3	1.2	0.90	0.16	0.46	1.0
0.1	1.4	2.1	1.7	1.4	0.86	1.4	2.3
0.3	2.7	3.5	2.4	–	1.1	3.2	3.1
1	4.8	–	3.1	–	1.4	6.2	4.5
3	8.6	–	4.0	–	1.8	10	7.7
10	22	–	11	–	6.0	19	18
30	72	–	26	–	18	37	42
100	–	–	74	–	53	130	100

Table 7.15 Transport coefficients for electrons in helium in the regime of a low electron number density. The reduced electric field strength E/N_a is expressed in Td (1 Td $= 10^{-17}$ V cm^2), the electron drift velocity w_e is given in cm/s, the reduced transversal diffusion coefficient of electrons, $D_\perp/N_a$, is represented in (cm s)$^{-1}$. The electron mobility $K = w_e/E$ is taken from [164, 178, 186], and the Townsend characteristic energy $eD_\perp/K_e$ is measured in eV. The gas temperature is 300 K

E/N_a	w_e	$D_\perp N_a$	$eD_\perp/K$
0.001	$2.5 \cdot 10^3$	$6.2 \cdot 10^{21}$	0.025
0.003	$7.4 \cdot 10^3$	$6.2 \cdot 10^{21}$	0.025
0.01	$2.5 \cdot 10^4$	$7.2 \cdot 10^{21}$	0.029
0.03	$7.4 \cdot 10^4$	$8.9 \cdot 10^{21}$	0.036
0.1	$1.6 \cdot 10^5$ ($1.7 \cdot 10^5$)	$1.0 \cdot 10^{22}$	0.064
0.3	$2.9 \cdot 10^5$ ($2.9 \cdot 10^5$)	$1.4 \cdot 10^{22}$	0.15
1	$4.9 \cdot 10^5$ ($5.3 \cdot 10^5$)	$2.2 \cdot 10^{22}$	0.44
3	$9.0 \cdot 10^5$ ($9.1 \cdot 10^5$)	$3.6 \cdot 10^{22}$	1.2
10	$2.1 \cdot 10^6$ ($1.7 \cdot 10^6$)	$7.6 \cdot 10^{22}$	3.6
30	$7.0 \cdot 10^6$ [$6.5 \cdot 10^6$]	$1.2 \cdot 10^{23}$	5.3
100	$2.50 \cdot 10^7$ [$2.2 \cdot 10^7$]	$2.2 \cdot 10^{23}$	8.2
300	$7.0 \cdot 10^7$ [$6.5 \cdot 10^7$]	$3.0 \cdot 10^{23}$	11

Table 7.16 Transport coefficients for electrons in helium in the regime of a low electron number density. The reduced electric field strength E/N_a is expressed in Td (1 Td $= 10^{-17}$ V cm^2), the electron drift velocity w_e is given in cm/s, the reduced transversal diffusion coefficient of electrons, $D_\perp/N_a$, is represented in (cm s)$^{-1}$. The electron mobility $K = w_e/E$ is taken from [178, 188–190], and the Townsend characteristic energy $eD_\perp/K_e$ is measured in eV. The gas temperature is 300 K

E/N_a, Td	w_e, 10^5 cm/s	$D_\perp N_a$, 10^{22} (cm s)$^{-1}$	$eD_\perp/K$, eV
0.001	0.15	5.7	0.038
0.003	0.48	21	0.13
0.01	1.0	47	0.47
0.03	1.4	47	1.0
0.1	1.8	29	1.6
0.3	2.3	19	2.5
1	2.7	12	4.4
3	4.3	10	6.9
10	10	7.1	7.1
30	24	5.9	7.4
100	60	5.0	8.3
300	140	4.0	8.5

Chapter 8
Conclusion

This book is devoted to properties of atomic particles, as well as to processes involving atomic particles. Basic models for parameters and dynamics of atomic particles are accompanied by numerical data in their tables and figures. The material of the book is useful for the numerical analysis of certain atomic objects and allows one to find or to estimate parameters of atomic objects as well as those for processes with participation of atomic particles. The data and models under consideration are of interest not only for atomic and molecular physics, but also to physics of gases, plasmas, liquids and solids, physical kinetics and thermodynamics where atomic interactions are essential. Hence, this book may be an addition to traditional courses related to various fields of physics where atomic interactions or atomic transitions are of importance.

© Springer International Publishing AG, part of Springer Nature 2018   183
B. M. Smirnov, *Atomic Particles and Atom Systems*, Springer Series on Atomic,
Optical, and Plasma Physics 51, https://doi.org/10.1007/978-3-319-75405-5_8

Appendix A
Spectra and Diagrams

Atomic Spectra

Physical Quantities in the Form of Periodical Tables.

Pt1. Standard weights of elements and their natural occurrence in Earth crust (Fig. 2.1).
    Pt2. Natural occurrence of stable isotopes (Fig. 2.2).
    Pt3. Long-lived radioactive isotopes (Fig. 2.3).
    Pt4. Radiactive isotopes of lanthanides and transuranides (Fig. 2.4).
    Pt5. Ionization potentials of atoms and ions (Fig. 3.19).
    Pt6. Ionization potentials of atoms and ions of lanthanides and transuranides (Fig. 3.20).
    Pt7. Electron affinities of atoms (Fig. 3.21).
    Pt8. Lowest excited states of atoms (Fig. 3.22).
    Pt9. Splitting of lowest atom levels (Fig. 3.23).
    Pt10. Resonantly excited atom states (Fig. 4.1).
    Pt11. Atomic and diatomic polarizabilities (Fig. 5.2).
    Pt12. Parameters of homonuclear molecules (Fig. 5.5).
    Pt13. Parameters of homonuclear positive diatomic ions (Fig. 5.6).
    Pt14. Parameters of homonuclear negative diatomic ions (Fig. 5.7).
    Pt15. Electron affinity of molecules (Fig. 5.18).
    Pt16. Affinity of atoms to hydrogen and oxygen atoms (Fig. 5.22).
    Pt17. Cross sections of resonant charge exchange (Fig. 6.4).
    Pt18. Parameters of cross section of resonant charge exchange (Fig. 6.5).
    Pt19. Mobilities of atomic ions in parent gas (Fig. 7.2).

© Springer International Publishing AG, part of Springer Nature 2018
B. M. Smirnov, *Atomic Particles and Atom Systems*, Springer Series on Atomic,
Optical, and Plasma Physics 51, https://doi.org/10.1007/978-3-319-75405-5

Figures

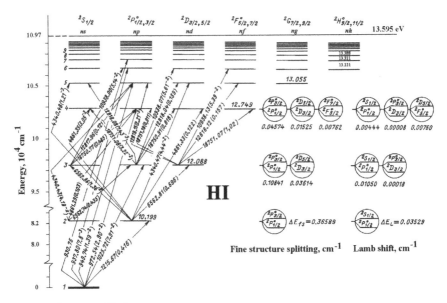

Fig. A.1 Spectrum of hydrogen atom

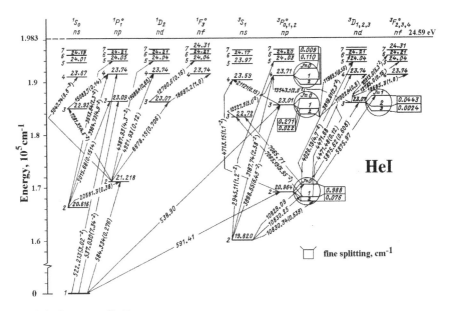

Fig. A.2 Spectrum of helium atom

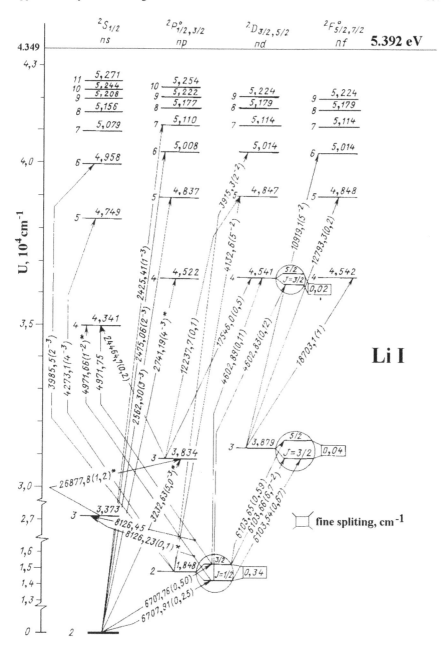

Fig. A.3 Spectrum of lithium atom

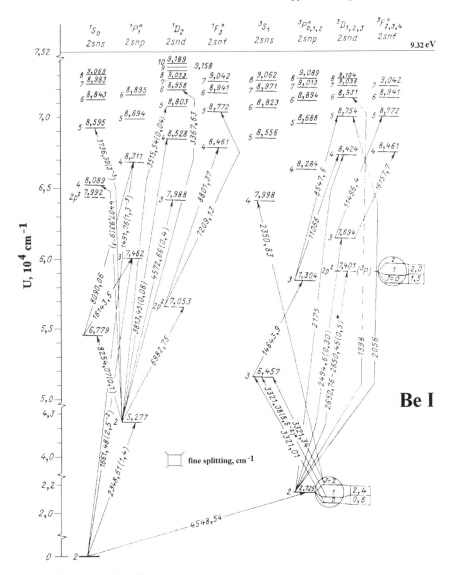

Fig. A.4 Spectrum of beryllium atom

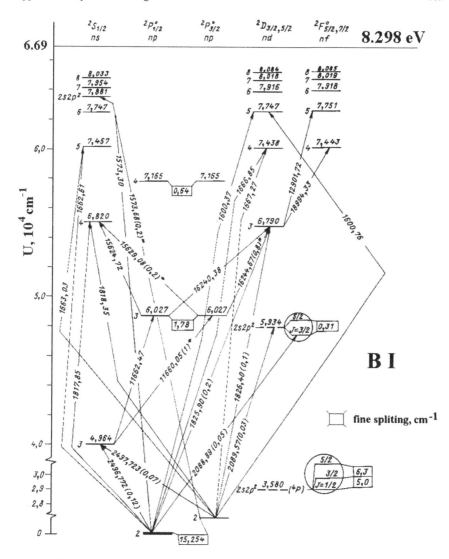

Fig. A.5 Spectrum of boron atom

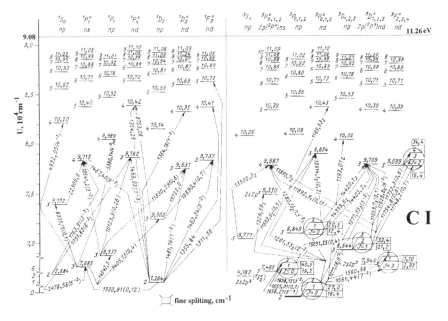

Fig. A.6 Spectrum of carbon atom

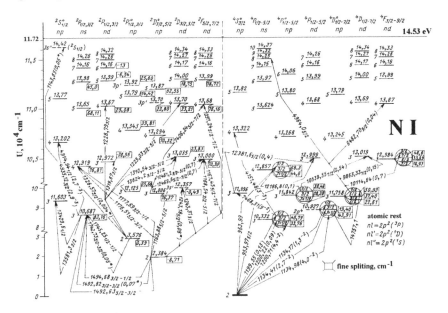

Fig. A.7 Spectrum of nitrogen atom

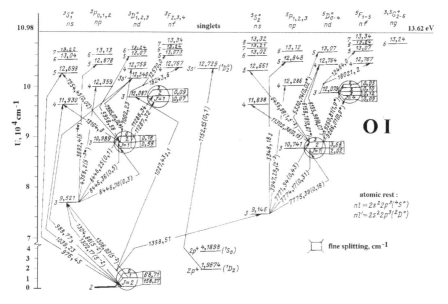

Fig. A.8 Spectrum of oxygen atom

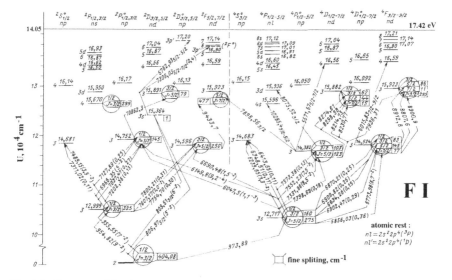

Fig. A.9 Spectrum of fluorine atom

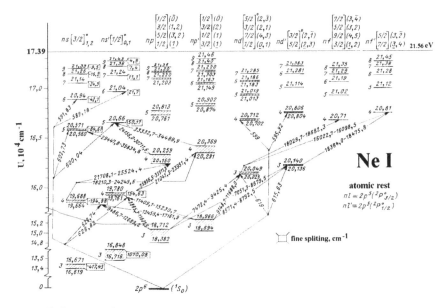

Fig. A.10 Spectrum of neon atom

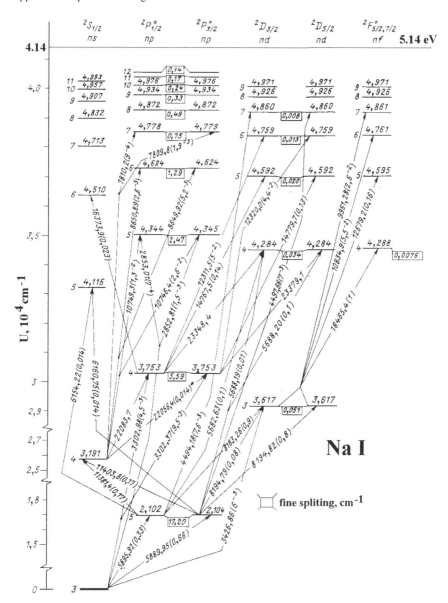

Fig. A.11 Spectrum of sodium atom

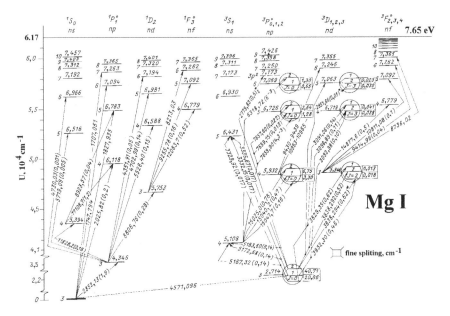

Fig. A.12 Spectrum of magnesium atom

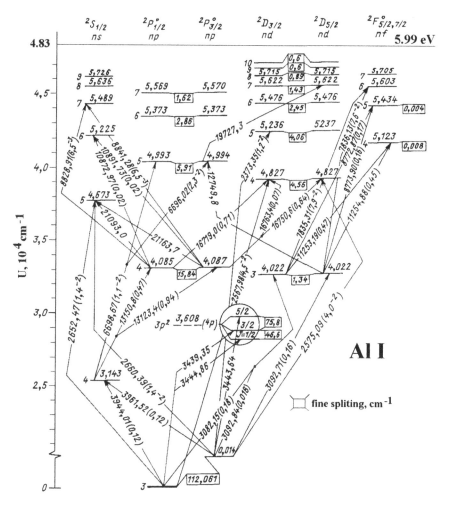

Fig. A.13 Spectrum of aluminium atom

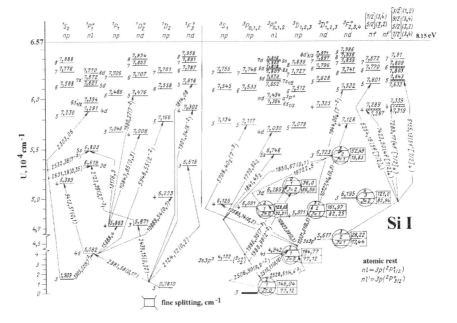

Fig. A.14 Spectrum of silicon atom

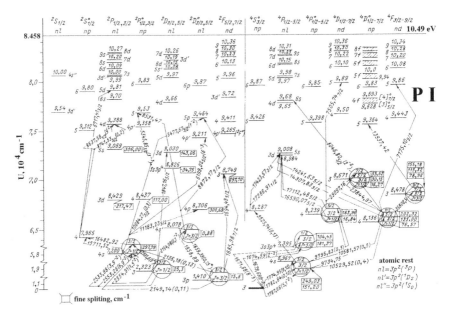

Fig. A.15 Spectrum of phosphorus atom

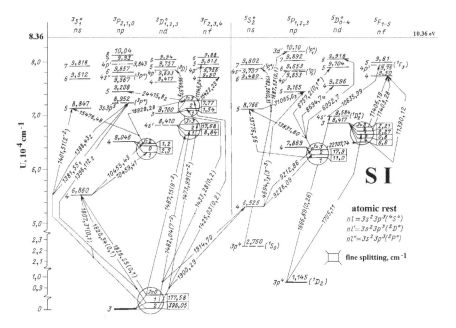

Fig. A.16 Spectrum of sulfur atom

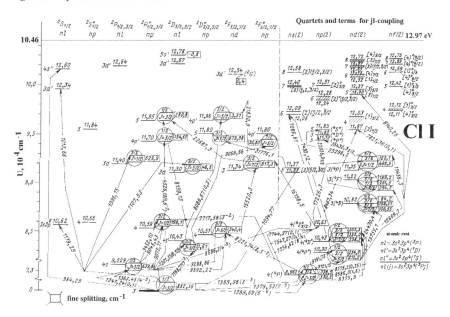

Fig. A.17 Spectrum of chlorine atom

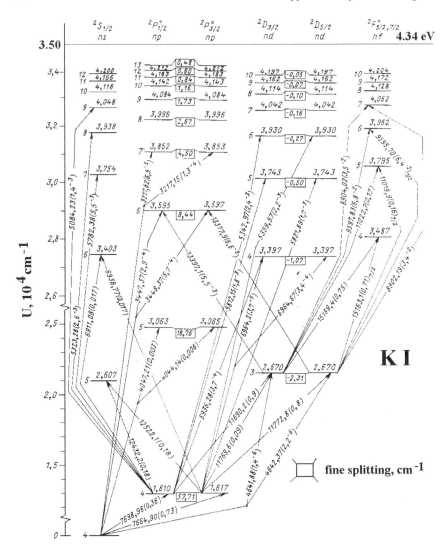

Fig. A.18 Spectrum of potassium atom

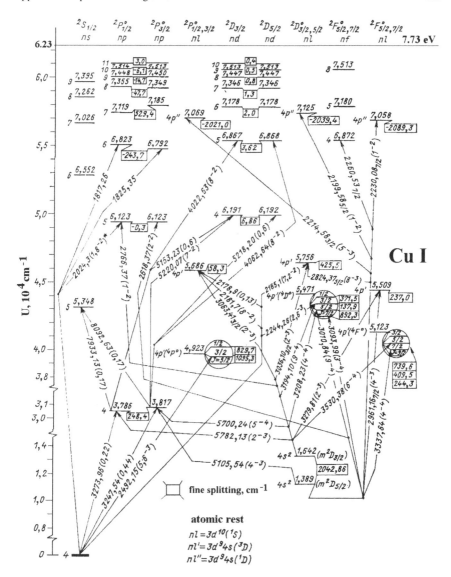

Fig. A.19 Spectrum of copper atom

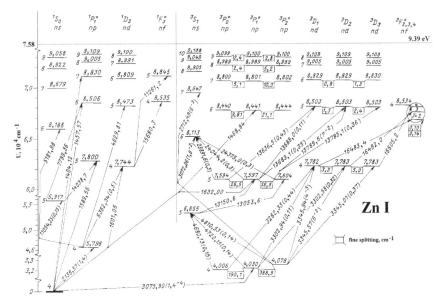

Fig. A.20 Spectrum of zinc atom

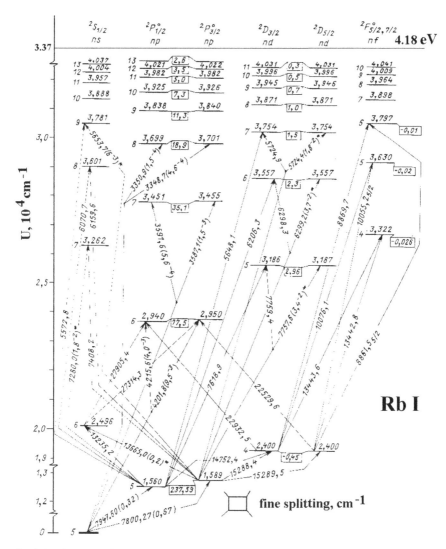

Fig. A.21 Spectrum of rubidium atom

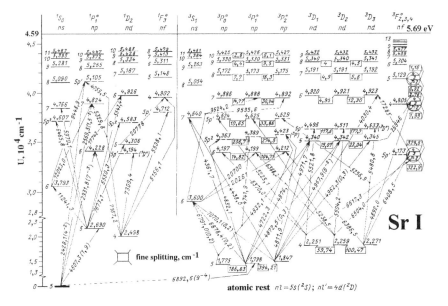

Fig. A.22 Spectrum of strontium atom

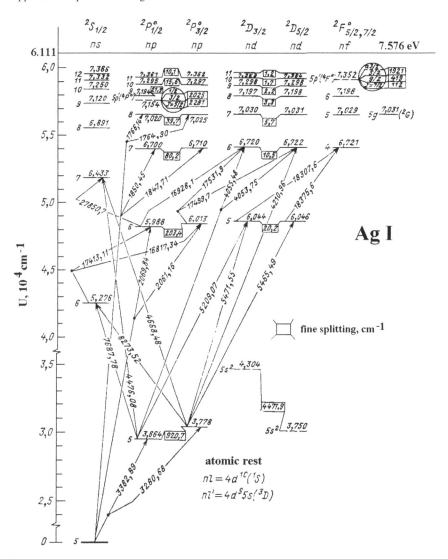

Fig. A.23 Spectrum of silver atom

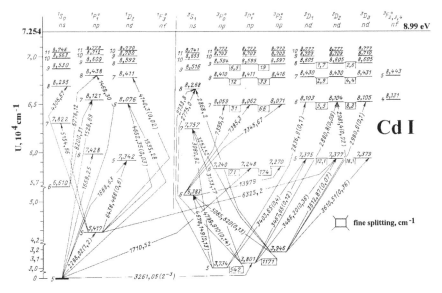

Fig. A.24 Spectrum of cadmium atom

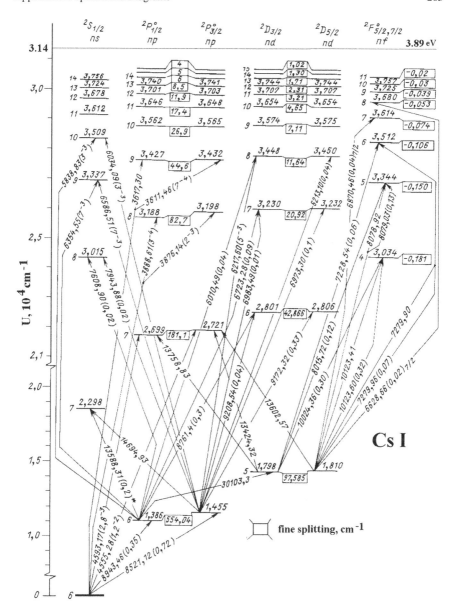

Fig. A.25 Spectrum of caesium atom

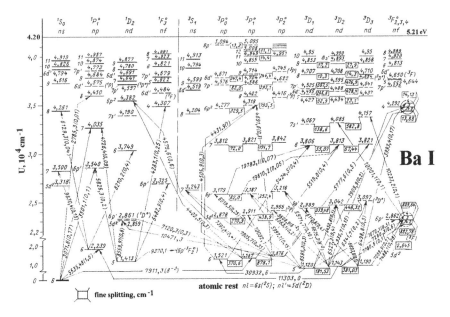

Fig. A.26 Spectrum of barium atom

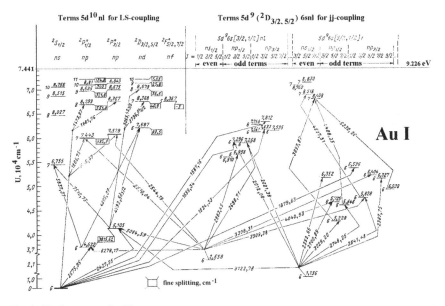

Fig. A.27 Spectrum of gold atom

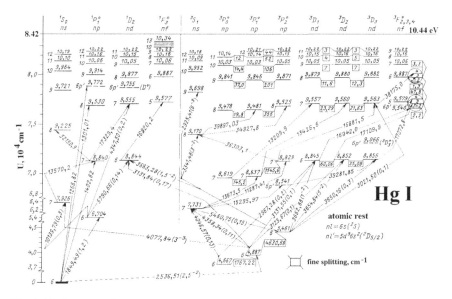

Fig. A.28 Spectrum of mercury atom

References

1. L.A. Sena, *Units of Physical Units and Their Dimensionalities* (Mir, Moscow, 1988)
2. W.M. Haynes (ed.), *CRC Handbook of Chemistry and Physics*, 97 edn. (CRC Press, London, 2016–2017)
3. http://en.wikipedia.org/wiki/Centimetre-gram-second-system-of-units
4. http://physics.nist.gov/cuu/units
5. D.E. Gray (ed.), *American Institute of Physics Handbook* (McGrow Hill, New York, 1971)
6. A.J. Moses, *The Practicing Scientists Book* (Van Nostrand, New York, 1978)
7. K.P. Huber, G. Herzberg, *Molecular Spectra and Molecular Structure, Constants of Diatomic Molecules* (Van Nostrand, New York, 1979)
8. A.A. Radzig, B.M. Smirnov, *Reference Data on Atoms, Molecules and Ions* (Springer, Berlin, 1985)
9. R.C. Reid, J.M. Prausnitz, B.E. Poling, *The Properties of Gases and Liquids*, 4th edn. (McGraw Hill, New York, 1987)
10. V.A. Rabinovich, *Thermophysical Properties of Neon, Argon, Krypton and Xenon* (Hemisphere, New York, 1987)
11. J. Emsley, *The Elements*, 2nd edn. (Claredon Press, Oxford, 1991)
12. L.V. Gurvich, V.S. Iorish et al., *IVTANTHERMO-A Thermodynamic Database and Software System for the Personal Computer, User's Guide* (CRC Press, Boca Raton, 1993)
13. S.V. Khristenko, A.I. Maslov, V.P. Shevelko, *Molecules and Their Spectroscopic Properties* (Springer, Berlin, 1998)
14. I. Barin, *Thermophysical Data of Pure Substances* (Wiley, New York, 2004)
15. S. Bashkin, J. Stoner, *Atomic Energy Levels and Grotrian Diagrams*, vols. 1–4 (Amsterdam, North Holland, 1975–1982)
16. A.S. Yatzenko, *Grotrian Diagrams of Neutral Atoms* (Nauka, Novosibirsk, 1993). (in Russian)
17. K.N. Huang, *Atomic Energy Levels and Grotrian Diagrams of Iron* (North Holland, Amsterdam, 1999)
18. G.G. Telegin, A.S. Yatzenko, *Optical Spectra of Atmospheric Gases* (Nauka, Novosibirsk, 2000). (in Russian)
19. B.M. Smirnov, *Physics of Atoms and Ions* (Springer, New York, 2003)
20. G. Herzberg, *Atomic Spectra and Atomic Structure* (Dover, New York, 1944)
21. H.A. Bethe, *Intermediate Quantum Mechanics* (Benjamin Inc., New York, 1964)
22. A. Schebline, P.B. Farnsworth, Spectrom. Acta **488**, 99 (1993)
23. https://www.nist.gov/srd
24. https://www.nist.gov/pml/productservices/physical-reference-data
25. https://www.nist.gov/pml/atomic-spectra-database
26. P.J. Mohr, D.B. Newell, B.N. Taylor, Rev. Mod. Phys. **88**, 035009 (2016)
27. L.S. Rothman e.a., J. Quant. Spectr. Rad. Trans. **60**, 665(1998)
28. L.S. Rothman e.a., J. Quant. Spectr. Rad. Trans. **110**, 533(2009)

29. L.S. Rothman, J. Quant. Spectr. Rad. Trans. **111**, 1565 (2010)
30. http://en.wikpedia.org/wiki/speed-of-light
31. http://en.wikpedia.org/wiki/Planck-constant
32. B.M. Smirnov, *Clusters and Small Particles in Gases and Plasmas* (Springer, New York, 2000)
33. B.M. Smirnov, *Physics of Ionized Gases* (Wiley, New York, 2001)
34. B.M. Smirnov, *Principles of Statistical Physics* (Wiley, Weinheim, 2006)
35. B.M. Smirnov, *Plasma Processes and Plasma Kinetics* (Wiley, Berlin, 2007)
36. B.M. Smirnov, *Cluster Processes in Gases and Plasmas* (Wiley, Berlin, 2010)
37. B.M. Smirnov, *Nanoclusters and Microparticles in Gases and Vapors* (de Gruyter, Berlin, 2012)
38. B.M. Smirnov, *Fundamental of Ionized Gases, Basic Topics in Ionized Gases* (Wiley, Weinheim, 2012)
39. V.Yu. Baranov (ed.), *Isotopes: Properties, Generation, Applications* (AtomIzdat, Moscow, 2000). (in Russian)
40. R.B. Firestone, *Tables of Isotopes* (Wiley, New York, 1996)
41. http://en.wikpedia.org/wiki/stableisotope
42. H. Budzikiewicz, R.D. Grigsby, Mass Spectrom. Rev. **25**, 146 (2006)
43. I.Ju. Tolstikhina, V.P. Shevelko, Phys. Uspekhi (2017)
44. H.A. Bethe, *Quantenmechanik der Ein-und Zweielectronenprobleme*. Handbuch der Physik, vol. 24–1 (Springer, Berlin, 1933)
45. H.A. Bethe, E.E. Salpeter, *Quantum Mechanics of One and Two-Electron Atoms* (Springer, Berlin, 1957)
46. L.D. Landau, E.M. Lifshitz, *Quantum Mechanics* (Pergamon Press, Oxford, 1980)
47. T.F. Gallagher, *Rydberg Atoms* (Cambridge University Press, Cambridge, 1994)
48. B. Edlen, *in Encyclopedia of Physics* (Springer, Berlin, 1964)
49. M. Seaton, Rep. Prog. Phys. **46**, 167 (1983)
50. V.S. Lebedev, I.L. Beigman, *Physics of Highly Excited Atoms and Ions* (Springer, Berlin, 1988)
51. V.S. Lebedev, I.L. Beigman, Phys. Rep. **250**, 95 (1995)
52. W. Pauli, Zs. Phys. **31**, 765 (1925)
53. W. Pauli, *Exclusion Principle and Quantum Mechanics* (Editions du Griffon, Neuchatel, Switzerland, 1947)
54. I.G. Kaplan, *The Pauli Exclusion Principle* (Wiley, Chichester, 2017)
55. G. Racah, Phys. Rev. **61**, 186 (1942); **62**, 438 (1942)
56. E.U. Condon, G.H. Shortley, *Theory of Atomic Spectra* (Cambridge University Press, Cambridge, 1949)
57. L.I. Sobelman, *Atomic Spectra and Radiative Transitions* (Springer, Berlin, 1979)
58. R.S. Mulliken, Rev. Mod. Phys. **2**, 60 (1930)
59. F. Hund, Zs. Phys. **36**, 637 (1936)
60. M.G. Veselov, L.N. Labtzovskij, *Theory of the Atom Structure of Electron Shells* (Nauka, Moscow, 1986). (in Russian)
61. B.M. Smirnov, Phys. Uspekhi **44**, 221 (2001)
62. W.C. Martin, R. Zalubas, Phys. Chem. Ref. Data **12**, 323 (1983)
63. F. Hund, *Linienspectren and periodisches System der Elemente* (Springer, Berlin, 1927)
64. J. Barrett, *Atomic Structure and Periodicity* (Wiley, Berlin, 2003)
65. V.P. Krainov, H.R. Reiss, B.M. Smirnov, *Radiative Processes in Atomic Physics* (Wiley, New York, 1997)
66. P.F. Bernath, *Spectroscopy of Atoms and Molecules* (Oxford University Press, Oxford, 2005)
67. W. Demdröter, *Atoms, Molecules and Photons* (Springer, Berlin, 2006)
68. W. Demdröter, *Laser Spectroscopy, Basic Principles* (Springer, Berlin, 2008)
69. S. Svanberg, *Atomic and Molecular Spectroscopy* (Springer, Berlin, 2012)
70. R. Kakkar, *Atomic and Molecular Spectroscopy: Basic Concepts and Applications* (Cambridge University Press, Delhi, 2015)

71. A.S. Yatzenko, *Grotrian Diagrams of Singly Charged Ions* (Nauka, Novosibirsk, 1996). (in Russian)
72. T. Shirai, J. Sugar, W. Wiese e.a., *Spectral Data and Grotrian Diagrams for Highly Ionized Atoms* (AIP, New York, 2000)
73. A.S. Yatzenko, *Grotrian Diagrams of Multicharged Ions* (Nauka, Novosibirsk, 2001). (in Russian)
74. A.S. Yatzenko, *Optical Spectra of H and He-like Ions* (Nauka, Novosibirsk, 2003). (in Russian)
75. A.S. Yatzenko, *Optical Spectra of Li and Be-like Ions* (Nauka, Novosibirsk, 2005). (in Russian)
76. V.S. Lisitsa, *Atoms in Plasmas* (Springer, Berlin, 1994)
77. W.L. Wiese, M.W. Smith, B.M. Glennon, *Atomic Transition Probabilities—H Through Ne*, vol. 1. National Standard Reference Data Series, vol. 4 (National Bureau of Standards, New York, 1966)
78. W.L. Wiese, M.W. Smith, B.M. Miles, *Atomic Transition Probabilities—Na Through Ca*, vol. 2. National Standard Reference Data Series, vol. 22 (National Bureau of Standards, New York, 1969)
79. W.L. Wiese, B.M. Glennon, Atomic transition probabilities, in *American Institute of Physics Handbook*, ed. by D.E. Gray (McGrow Hill, New York, 1972), pp. 200–263, Chap. 7
80. W.L. Wiese, G.A. Martin, Transition probabilities, in *Wavelengths and Transition Probabilities for Atoms and Atomic Ions, Part II*. National Standard Reference Data Series, vol. 68 (National Bureau of Standards, New York, 1980), pp. 359–406
81. http://physics.nist.gov/PhysRefData/ASD1
82. L.A. Vainstein, I.I. Sobelman, E.A. Yukov, *Excitation of Atoms and Broadening of Spectral Lines* (Nauka, Moscow, 1979). (in Russian)
83. J.H. Lambert, *Photometry, or, On the Measure and Gradations of Light, Colors, and Shade* (Augsburg, Eberhardt Klett, 1760)
84. A. Beer, Annalen der Physik und. Chemie. **86**, 78 (1852)
85. W. Heitler, F. London, Phys. Zs. **44**, 445 (1927)
86. S.C. Wangm, Phys. Zs. **28**, 363 (1927)
87. A. Dalgarno, Rev. Mod. Phys. **35**, 522 (1963)
88. A.E. Kingston, Phys. Rev. **135A**, 1018 (1964)
89. A. Dalgarno, Adv. Chem. Phys. **12**, 143 (1967)
90. E.E. Fermi, Nuovo cim. **11**, 157 (1934)
91. M.Ya. Ovchinnikova, Sov. Phys. JETP **22**, 194 (1966)
92. B.M. Smirnov, Teor. Exsp. Khim. **22**, 194 (1971)
93. V.N. YuN Demkov, Ostrovskii, *Method of Potentials of Zero-th Radius in Atomic Physics* (Plenum Press, New York, 1988)
94. C. Herzberg, *Molecular Spectra and Molecular Structure*, vol. 1–4 (van Nostrand, New York, 1964–1966)
95. J.M. Brown, *Molecular Spectroscopy* (Oxford University Press, Oxford, 1998)
96. E.E. Nikitin, Optika Spectr. **22**, 379 (1966)
97. E.E. Nikitin, B.M. Smirnov, Sov. Phys. Uspekhi **21**, 95 (1978)
98. E.E. Nikitin, S.J. Umanskii, *Theory of Slow Atomic Collisions* (Springer, Berlin, 1984)
99. E.E. Nikitin, B.M. Smirnov, *Atomic and Molecular Processes* (Nauka, Moscow, 1988). (in Russian)
100. B.M. Smirnov, in *Theory of Chemical Reaction Dynamics*, ed. by A. Lagana, G. Landvay. NATO Series, vol. 145 (2004), pp. 129–148
101. L.D. Laude (ed.), *Excimer Lasers* (NATO Series E, Applied Sciences (Dordrecht, Kluwer, 1989)
102. L.D. Laude, *Excimer Lasers* (Springer, Berlin, 1994)
103. D. Basting, G. Marowsky, *Excimer Laser Technology* (Springer, Berlin, 2005)
104. E. Wigner, E.E. Witmer, Zs. Phys. **51**, 859 (1928)
105. R.S. Mulliken, Rev. Mod. Phys. **2**, 1440 (1930)
106. R.S. Mulliken, Rev. Mod. Phys. **4**, 1 (1932)

107. G. Herzberg, *Molecular Spectra and Molecular Structure. I. Diatomic Molecules* (New York, 1939)
108. A.D. Gaydon, *Dissociative Energies and Spectra of Diatomic Molecules* (London, 1947)
109. L.B. Loeb, *Basic Processes of Gaseous Electronics* (University of California Press, Berkeley, 1955)
110. N.F. Mott, H.S.W. Massey, *The Theory of Atomic Collisions* (Claredon Press, Oxford, 1965)
111. E.W. McDaniel, J.B.A. Mitchell, M.E. Rudd, *Atomic Collisions* (Wiley, New York, 1993)
112. L.D. Landau, E.M. Lifshitz, *Mechanics* (Pergamon Press, London, 1980)
113. G.F. Drukarev, *Collissions of Electrons with Atoms and Molecules* (Plenum Press, New York, 1987)
114. P. Burke, *Theory of Electron-Atom Collisions* (Cambridge University Press, Cambridge, 1995)
115. H.S.W. Massey, E.H.S. Burhop, *Electronic and Ionic Impact Phenomena* (Oxford University Press, Oxford, 1969)
116. A. Dalgarno, Philos. Trans. **A250**, 428 (1958)
117. E.W. McDaniel, E.A. Mason, *The Mobility and Diffusion of Ions in Gases* (Wiley, New York, 1973)
118. H.S.W. Massey, *Atomic and Molecular Collisions* (Taylor and Francis, New York, 1979)
119. E.A. Mason, E.W. McDaniel, *Transport Properties of Ions in Gases* (Wiley, New York, 1988)
120. O.B. Firsov, Zh Exp, Theor. Fiz. **21**, 1001 (1951)
121. D. Rapp, W.E. Francis, J. Chem. Phys. **37**, 2631 (1962)
122. H.S.W. Massey, H.B. Gilbody, *Electronic and Ionic Impact Phenomena* (Oxford University Press, London, 1974)
123. S. Sakabe, Y. Izawa, Atom. Data Nucl. Data Tabl. **49**, 257 (1991)
124. L.A. Sena, Zh Exp, Theor. Fiz. **9**, 1320 (1939)
125. L.A. Sena, Zh Exp, Theor. Fiz. **16**, 734 (1946)
126. L.A. Sena, *Collisions of Electrons and Ions with Atoms* (Gostekhizdat, Leningrad, 1948). (in Russian)
127. B.M. Smirnov, Sov. Phys. JETP **19**, 692 (1964)
128. B.M. Smirnov, Sov. Phys. JETP **20**, 345 (1965)
129. B.M. Smirnov, Phys. Scripta **61**, 595 (2000)
130. A.V. Kosarim, B.M. Smirnov, JETP **101**, 611 (2005)
131. A.V. Kosarim, B.M. Smirnov, M. Capitelli, R. Celiberto, A. Larichuita, Phys. Rev. **74A**, 062707 (2006)
132. B.M. Smirnov, Asymptotic theory of charge exchange and mobility processes for atomic ions, in *Reviews of Plasma Physics*, vol. 23, ed. by V.D. Shafranov (Kluwer Academic, New York, 2013)
133. B.M. Smirnov, JETP **92**, 951 (2001)
134. B.M. Smirnov, *Ions and Excited Atoms in Plasma* (Atomizdat, Moscow, 1974). (in Russian)
135. B.M. Smirnov, *Excited Atoms* (Energoatomizdat, Moscow, 1982). (in Russian)
136. Y. Faxen, J. Holtzmark, Zs. Phys. **45**, 307 (1927)
137. H. Bethe, Phys. Rev. **76**, 38 (1949)
138. J.L. Pack e.a., J. Appl. Phys. **71**, 5363 (1992)
139. K. Tashibana, Phys. Rev. A **34**, 1007 (1986)
140. www.lxcat.net/Phelps
141. C. Ramsauer, Ann. Physik **72**, 345 (1923)
142. C. Ramsauer, R. Kollath, Ann. Physik **3**, 536 (1929)
143. L.S. Frost, A.V. Phelps, Phys. Rev. **136A**, 1538 (1964)
144. T. Koizumi, E. Shirakawa, E. Ogawa, J. Phys. B **19**, 2334 (1986)
145. D.F. Register, L. Vuskovic, S. Trajimar, J. Phys. B **19**, 1685 (1986)
146. R.P. McEachran, A.D. Stauffer, J. Phys. B **20**, 3483 (1987)
147. M. Suzuki et al., J. Phys. **D25**, 50 (1992)
148. T.F. O'Malley, L. Spruch, T. Rosenberg, J. Math. Phys. **2**, 491 (1961)
149. T.F. O'Malley, Phys. Rev. **130**, 1020 (1963)
150. B.M. Smirnov, *Physics of Ionized Gases* (Atomizdat, Moscow, 1972). (in Russian)

151. B.M. Smirnov, *Atomic Collisions and Elementary Processes in Plasma* (Atomizdat, Moscow, 1968). (in Russian)
152. M. Capitelli, C.M. Ferreira, B.F. Gordiets, A.I. Osipov, *Plasma Kinetics in Atmospheric Gases* (Springer, Berlin, 2000)
153. B.M. Smirnov, *Microphysics of Atmospheric Phenomena* (Springer, Switzerland, 2017)
154. S. Chapman, T.G. Cowling, *The Mathematical Theory of Non-uniform Gases* (Cambridge University Press, Cambridge, 1952)
155. J.H. Ferziger, H.G. Kaper, *Mathematical Theory of Transport Processes in Gases* (North Holland, Amsterdam, 1972)
156. M. Capitelli, D. Bruno, A. Laricchiuta, *Fundamental Aspects of Plasma Chemical Physics* (Springer, New York, 2013)
157. N.B. Vargaftik, *Tables of Thermophysical Properties of Liquids and Gases* (Halsted Press, New York, 1975)
158. C.J. Zwakhals, K.W. Reus, Physica C **100**, 231 (1980)
159. *Reference Data for Rate Constants of Elementary Processes involving Atoms, Ions, Electrons and Photons* (Izd. Petersb. Univ., Petersburg, 1994) (in Russian)
160. N.B. Vargaftic, L.N. Filipov, A.A. Tarismanov, E.E. Totzkii, *Reference Book on Thermal Properties of Liquids and Gases* (CRC Press, Boca Raton, 1994)
161. A. Einstein, Ann. Phys. **17**, 549 (1905)
162. A. Einstein, Ann. Phys. **19**, 371 (1906)
163. J.S. Townsend, V.A. Bailey, Phil. Trans. A **193**, 129 (1899); **195**, 259 (1900)
164. L.G.H. Huxley, R.W. Crompton, *The Diffusion and Drift of Electrons in Gases* (Wiley, New York, 1973)
165. A. Dalgarno, M.R.C. McDowell, A. Williams, Phil. Transact. A **250**, 411 (1958)
166. A.M. Tyndall, *The Mobility of Positive Ions in Gases* (Cambridge University Press, Cambridge, 1938)
167. R.W. Crompton, M.T. Elford, Proc. Roy. Soc. **74**, 497 (1964)
168. H.W. Ellis, R.Y. Pai, E.W. McDaniel, E.A. Mason, L.A. Viehland, Atomic Data Nucl. Data Tabl. **17**, 177 (1976)
169. H.W. Ellis, E.W. McDaniel, D.L. Albritton, L.A. Viehland, S.L. Lin, E.A. Mason, Atomic Data Nucl. Data Tabl. **22**, 179 (1978)
170. H.W. Ellis, M.G. Trackston, E.W. McDaniel, E.A. Mason, Atomic Data Nucl. Data Tabl. **31**, 113 (1984)
171. L.A. Viehland, E.A. Mason, Atom. Data Nucl. Data Tabl. **60**, 37 (1995)
172. T. Holstein, J. Chem. Phys. **56**, 832 (1952)
173. B.M. YuP Mordvinov, Smirnov, Zh Eksp, Teor. Fiz. **48**, 133 (1965)
174. T. Kihara, Rev. Mod. Phys. **25**, 944 (1953)
175. V.I. Perel, Zh Exp, Theor. Fiz. **32**, 526 (1957)
176. B.M. Smirnov, Doklady Akad. Nauk SSSR **181**, 61 (1968)
177. W.P. Allis, in *Handbuch der Physik*, vol. 21, ed. by S. Flugge (Springer, Berlin, 1956), p. 383
178. J. Dutton, J. Phys. Chem. Ref. Data **4**, 577 (1975)
179. J.L. Pack, R.E. Voshall, A.V. Phelps, Phys. Rev. **127**, 2084 (1962)
180. S.S.S. Huang, G.R. Freeman, J. Chem. Phys. **68**, 1355 (1978)
181. A.L. Gillardini, *Low-Electron Collisions in Gases* (Wiley, New York, 1972)
182. I.E. McCarthy, E. Weigold, *Electron-Atom Collisions* (Cambridge University Press, Cambridge, 1995)
183. C.W. McCutchen, Phys. Rev. **112**, 1848 (1958)
184. Y. Nakamura, J. Lucas, J. Phys. **D11**, 325 (1978)
185. J.S. Townsend, V.A. Bailey, Philos. Mag. **42**, 873 (1921)
186. H.N. Kücükarpaci, H.T. Saelee, J. Lucas, J. Phys. **14D**, 9 (1981)
187. B.M. Smirnov, *Theory of Gas Discharge Plasma* (Springer, Heidelberg, 2017)
188. C.S. Lakshminarasimha, J. Lucas, J. Phys. **10D**, 313 (1977)
189. A.G. Robertson, Aust. J. Phys. **30**, 391 (1977)
190. H.B. Millloy, R.W. Crompton, Aust. J. Phys. **30**, 51 (1977)

Index

© Springer International Publishing AG, part of Springer Nature 2018
B. M. Smirnov, *Atomic Particles and Atom Systems*, Springer Series on Atomic,
Optical, and Plasma Physics 51, https://doi.org/10.1007/978-3-319-75405-5

Printed in the United States
By Bookmasters